Oxford Revision Guides

GCSE

Design and Technology

FOOD TECHNOLOGY
through diagrams

WITHDRAWN

Hazel King

OXFORD
UNIVERSITY PRESS

OXFORD
UNIVERSITY PRESS

Great Clarendon Street, Oxford OX2 6DP

Oxford University Press is a department of the
University of Oxford. It furthers the University's objective
of excellence in research, scholarship, and education by
publishing worldwide in

Oxford New York

Auckland Bangkok Buenos Aires
Cape Town Chennai Dar es Salaam Delhi
Hong Kong Instanbul Karachi Kolkata Kuala Lumpar
Madrid Melbourne Mexico City Mumbai Nairobi
São Paulo Shanghai Taipei Tokyo Toronto

Oxford is a registered trade mark of Oxford University Press
in the UK and in certain other countries

British Library Cataloguing in Publication Data

data available

ISBN 0 19 832817 6 (school edition)
     0 19 832818 4 (bookshop edition)
10 9 8 7 6 5 4

Typeset and designed by Hardlines, Charlbury. Oxon

Printed and bound in Great Britain

CONTENTS

Information in a shaded box represents knowledge of a higher level.

Syllabuses for GCSE Design and Technology : Food Technology

GCSE Food Technology comes under the heading of Design and Technology. It may be studied as a full or short course. Both courses can be assessed at a foundation or higher tier (level). The grades available for each tier are shown below:

	Tier	Grades available
GCSE Food Technology full and short courses	Higher	A* – D
	Foundation	C – G

EXAMINING GROUPS AND SYLLABUSES

There are several examining groups or boards that set syllabuses and examinations. Make sure you know which group and syllabus you are following. They may differ in coursework requirements, length of examination papers, etc. The content of all Food Technology syllabuses is broadly the same because they must follow the National Curriculum for Key Stage 4. Therefore this revision guide will help all GCSE Food Technology students, regardless of the syllabus being studied.

Examination group/syllabus	Course type	Tier	Length of paper	Coursework	Assessment		
NEAB Northern Examinations and Assessment Board	Full 1314	Higher	$2\frac{1}{2}$ hours	Design and Make Project 25 hours supervised time	Coursework	Exam	Total
		Foundation	2 hours		Designing 20% 20% 40% Making 40% 20% 60%		
	Short 2314	Higher	2 hours	Design and Make Project 12 hours supervised time			
		Foundation	$1\frac{1}{2}$ hours		Total 60% 40% 100%		
EDEXCEL London Examinations	Full 1966	Higher	$2\frac{1}{2}$ hours	i) Design Folder + practical outcome (30 hours) ii) Folio from investigation of a product (10 hours)	Coursework	Exam	Total
		Foundation	$2\frac{1}{2}$ hours		i) 40% 40% 100% ii) 20%		
	Short 3966	Higher	$1\frac{1}{2}$ hours	i) Design Folder + practical outcome (15 hours) ii) Investigation of a product (5 hours)			
		Foundation	$1\frac{1}{2}$ hours		Total 60% 40% 100%		
OCR (formally MEG, Midland Examining Group)	Full 1460	Higher	2 1 hour 15 mins 4 1 hour 15 mins	Design and Make Project Design folder and practical outcome (40–50 hours)	Coursework	Exam	Total
		Foundation	1 1 hour 3 1 hour		60% 40% 100%		
	Short 3460	Higher	2 1hour 15 mins	Design and Make Project Design folder and practical outcome (20–30 hours)	60% 40% 100%		
		Foundation	1 hour				
AQA (formerly SEG, Southern Examining Group)	Full 3400	Higher	3 $1\frac{1}{2}$ hours 1 1 hour	One integrated assignment to design and make a food product supported by second integrated assignment, activities or tasks (40 hours)	Coursework	Exam	Total
		Foundation	2 $1\frac{1}{2}$ hours 4 1 hour		60% 40% 100%		
	Short 1400	Higher	3 $1\frac{1}{2}$ hours	One integrated assignment + design and make a food product which may be supported by other activities or tasks (20 hours)			
		Foundation	2 $\frac{1}{2}$ hours				

Designing and making skills

GCSE Food Technology provides you with opportunities to use food for investigation, designing, making and evaluation. In order to develop designing and making skills you must ensure you have a sound knowledge and understanding of food as a material. It is important that you know about the various properties of food, the effect processing can have and which equipment and tools to choose in order to shape, cut, combine or finish a food product. With a thorough knowledge and understanding of food you will be able to produce a quality food outcome and a quality Food Technology project!

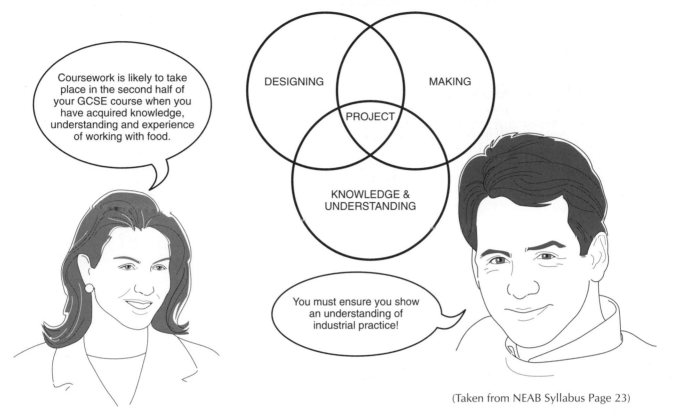

Coursework is likely to take place in the second half of your GCSE course when you have acquired knowledge, understanding and experience of working with food.

DESIGNING   MAKING

PROJECT

KNOWLEDGE & UNDERSTANDING

You must ensure you show an understanding of industrial practice!

(Taken from NEAB Syllabus Page 23)

COURSEWORK INFORMATION

Different syllabuses have different coursework requirements. Make sure you understand **exactly** what you are expected to do before you begin.

Group/ syllabus	Coursework requirements
NEAB	Design and Make Project consisting of a design folio and practical outcome. Making skills: 40%; designing skills: 20%. A project outline may be given to you or you may have a choice. It could be set by the examination board or by your teacher or you could write your own project outline but it must be approved by your teacher.
EDEXCEL London	One portfolio which includes (i) design folder and practical outcome(s) (ii) folio from an investigation of a product Full : (i) You have to complete a task in which you design and make food products. (ii) You will look into, disassemble and evaluate familiar food products or part products. Short : (i) You have to complete a project in which you design and make food products. (ii) You will review, disassemble and evaluate food products or part products. Both : You will choose your own theme/outline but it must be approved by your teacher.
OCR MEG	Full : 40–50 hours Short : 20–30 hours A project where you design and make a quality Food Technology product. The design and make activity must relate to industrial/commercial practices and the application of systems and control.
AQA SEG	Full : 40 hours An integrated assignment to design and make a product. This may be supported by a second integrated assignment or by other activities or tasks (eg evaluation of a product). Supporting activities may be part of, or go alongside, main assignment.	Short : 20 hours A substantial integrated assignment to design and make a product. This may be supported by other activities or task (eg develop and practice specific skills). Supporting activities may be part of, or go alongside, main assignment.

Preparing for revision

The key to getting through exams successfully is to be really prepared. Preparation begins with revision.

Exams? They're a piece of cake!

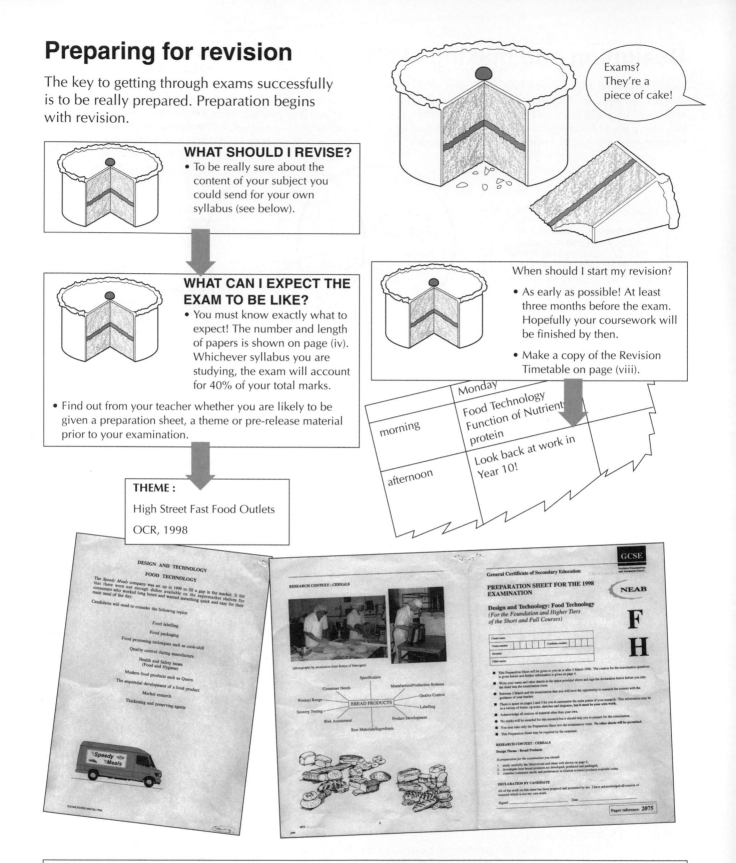

WHAT SHOULD I REVISE?
- To be really sure about the content of your subject you could send for your own syllabus (see below).

WHAT CAN I EXPECT THE EXAM TO BE LIKE?
- You must know exactly what to expect! The number and length of papers is shown on page (iv). Whichever syllabus you are studying, the exam will account for 40% of your total marks.

- Find out from your teacher whether you are likely to be given a preparation sheet, a theme or pre-release material prior to your examination.

When should I start my revision?
- As early as possible! At least three months before the exam. Hopefully your coursework will be finished by then.
- Make a copy of the Revision Timetable on page (viii).

	Monday
morning	Food Technology Function of Nutrients protein
afternoon	Look back at work in Year 10!

THEME :

High Street Fast Food Outlets

OCR, 1998

Syllabus contacts

- **NEAB**
  Northern Examinations and Assessment Board
  12 Harter Street, Manchester M1 6HL
  Tel: 0161 953 1170 Fax: 0161 953 1177

- **EDEXCEL**
  London Examinations
  Adamsway, Mansfield, Notts. NG18 4LN
  Tel: 01623 467467 Fax: 01623 450481

- **OCR**
  (formerly Midland Examining Group)
  Accounts Department
  OCR, Mill Wharf, Mill Street, Birmingham B6 4BU
  Tel: 01223 552552 Fax: 01223 553030

- **AQA** (SEG and AEB)
  Assessment and Qualifications Alliance
  SEG Publications Department
  Stag Hill House, Guildford, Surrey GU2 5XJ
  Tel: 01483 302302 Fax: 01483 300152

WHERE DO I BEGIN?

• Start by getting yourself organized. In order to revise thoroughly and effectively you will need the following:

HOW WILL THIS BOOK HELP ME?

• The information provided in this book covers the main content of Food Technology syllabuses. It is designed to remind you of topics you have already covered in your lessons. You may also discover details that are new to you. Presenting information in diagram form should help you to remember it.

SO WHY DON'T I JUST SIT AND READ THIS BOOK?

• Reading it is not enough! To revise and remember a topic you must be active!

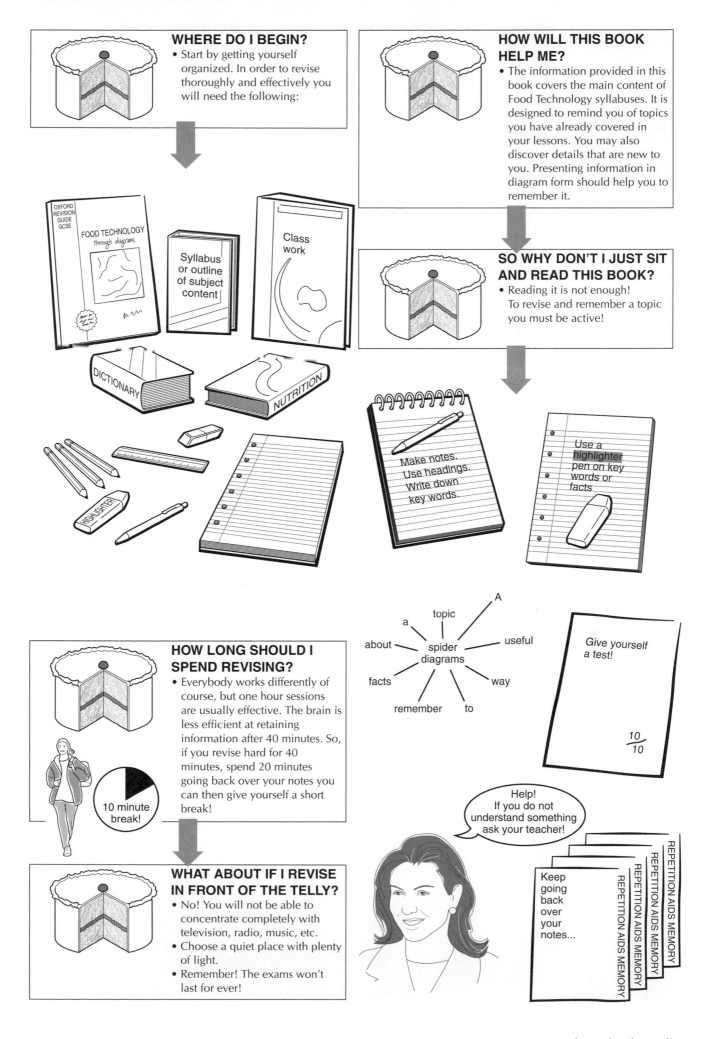

OXFORD REVISION GUIDE GCSE

FOOD TECHNOLOGY *through diagrams*

Syllabus or outline of subject content

Class work

DICTIONARY

NUTRITION

HIGHLIGHTER

Make notes. Use headings. Write down key words.

Use a highlighter pen on key words or facts

HOW LONG SHOULD I SPEND REVISING?

• Everybody works differently of course, but one hour sessions are usually effective. The brain is less efficient at retaining information after 40 minutes. So, if you revise hard for 40 minutes, spend 20 minutes going back over your notes you can then give yourself a short break!

10 minute break!

spider diagrams — A useful way to remember facts about a topic

Give yourself a test!

10/10

WHAT ABOUT IF I REVISE IN FRONT OF THE TELLY?

• No! You will not be able to concentrate completely with television, radio, music, etc.
• Choose a quiet place with plenty of light.
• Remember! The exams won't last for ever!

Help! If you do not understand something ask your teacher!

Keep going back over your notes...

REPETITION AIDS MEMORY
REPETITION AIDS MEMORY
REPETITION AIDS MEMORY
REPETITION AIDS MEMORY
REPETITION AIDS MEMORY

Revision Timetable

Week beginning: ....

Week no: .... Exam in: .... weeks!

Have you remembered to take a break?

Have you timetabled your favourite TV programme?

	Monday	Tuesday	Wednesday	Thursday	Friday	Saturday	Sunday	*notes*
morning								
afternoon								
evening (don't work too late!)								

End of day check: more work needed on?

1. ....

2. ....

3. ....

Timetable check: extra topics for next week?

1. ....

2. ....

3. ....

Success in the examination

The following advice may sound obvious but it is amazing how many candidates mis-read instructions or questions during an exam!

First, read through all instructions carefully.

Note how many questions you must answer, especially if the paper is divided into different sections.

Do not panic or rush. Calmly read through all the questions to give your brain a chance to start registering different topics.

Before answering a question, read it through a couple of times. Make sure you know **exactly** what you are being asked.

Start with a question you are confident about answering – this will help you to relax.

Look at the marks awarded to each question. Spend longer on those awarded higher marks.

Do not waste time over a question you find really difficult. You can always come back to it if you have time at the end.

If you do finish with time to spare, check over your answers. It is easy to misspell a word or even to use the wrong word when under pressure in an exam. Checking your answers could gain you some valuable marks.

EXAMINATION QUESTIONS

Different types of questions are used in examinations. Here are some examples of the type of questions you might be asked with suggested answers.

Short-answer questions

Typical example:
Name the country of origin of each of the three bread products in the table below:

Bread product	Country of origin
ciabatta	Italy
croissant	France
naan	India (Pakistan/Middle East also acceptable)

(3 marks)
(NEAB full course, foundation 1998)

Open-ended questions

Typical example:
The development of packaging has allowed a wide range of convenience foods to be sold e.g. cook/chill curry.

Explain how the packaging has enabled this to take place.

(2 marks)
(London full course, higher 1998)

Advances in technology have produced plastics suitable for the microwave, conventional oven and boil-in-the-bag products. Food can now be cooked in its packaging. Dual-purpose packaging makes products convenient to store and cook. The development of safety buttons and tamper-proof seals have also increased the packaging range and technology has enabled foods to be both preserved and packaged, eg asceptic packaging and modified atmosphere packaging.

STRUCTURED QUESTIONS

Typical example: The *Speedy Meals* company wishes to extend its range of international cook-chill products.

Explain what is meant by the term cook-chill.

Explanation **A dish which has been prepared, cooked, packaged and chilled rapidly or to 0–3 °C by the manufacturer which only needs re-heating by the consumer/microwaved.**

(2 marks)
(SEG foundation 1998)

Questions based on information provided

Typical example: Look at the packaging shown below and then answer the question opposite.

St Michael Homestyle

FRESH PARSNIP & GRUYERE BAKE

fresh parsnip slices in a rich cheese sauce with cheese crumble topping

SERVING SUGGESTION

SUITABLE FOR VEGETARIANS

SUITABLE FOR HOME FREEZING

READY TO COOK

300 g ℮ / 10.6 oz

DISPLAY UNTIL 02 DEC

USE BY 03 DEC

PRICE £1.69

KEEP REFRIGERATED 0°C to +5°C

INGREDIENTS

PARSNIP · CREAM · MILK · VEGETARIAN GRUYERE CHEESE · VEGETARIAN CHEDDAR CHEESE · BREADCRUMBS · SALT · SUGAR · MODIFIED STARCH · PEPPER.

NUTRITION

AVERAGE VALUES	PER 100g	PER 300g 10.6oz PACK
ENERGY	695 kJ	2085 kJ
	165 kcal	495 kcal
PROTEIN	4.9 g	14.7 g
CARBOHYDRATE	10.9 g	32.7 g
OF WHICH SUGARS	0.7 g	2.1 g
FAT	11.6 g	34.8 g
OF WHICH SATURATES	7.4 g	22.2 g
FIBRE	2.1 g	6.3 g
SODIUM	0.31 g	0.93 g

STORAGE

SUITABLE FOR HOME FREEZING
FREEZE ON DAY OF PURCHASE
USE WITHIN 1 MONTH
DEFROST THOROUGHLY
BEFORE MICROWAVING

COOKING

FOR BEST RESULTS: **CONVENTIONAL OVEN** - PREHEAT OVEN TO 190°C, 375°F, GAS 5, FAN OVEN 160°C. REMOVE CARTON AND LID. PLACE ON A PREHEATED BAKING TRAY IN OVEN FOR 30 MINUTES.

TO COOK FROM FROZEN - CONVENTIONAL - FOLLOW COOKING INSTRUCTIONS AND INCREASE COOKING TIME TO 40 MINUTES.

MICROWAVE OVEN - THE FOIL CONTAINER MUST NOT BE USED IN THE MICROWAVE OVEN AND THE PRODUCT SHOULD BE PLACED IN A NON-METALLIC DISH OF SIMILAR DIMENSIONS.
MICROWAVE OVENS VARY.
THE FOLLOWING RECOMMENDATION IS A GUIDE ONLY. PLACE IN OVEN AND COOK ON HIGH (100%).

CATEGORY	OR	WATTAGE	
B	D	650	750
3½ MINUTES	3 MINUTES	3½ MINUTES	3 MINUTES

ADJUST TIME ACCORDING TO YOUR PARTICULAR OVEN.

CHECK THAT PRODUCT IS HOT BEFORE SERVING.

TWO OR MORE PACKS WILL REQUIRE LONGER COOKING TIME. **DO NOT REHEAT.**

UK-LI 003   MADE IN THE U.K. · MARKS AND SPENCER p.l.c. · BAKER STREET · LONDON · ENGLAND © M&S 97   1813   TF 006 A

From the information on the packaging opposite, explain under the given headings, why consumers may be encouraged to buy the product.

Dietary needs

It is suitable for vegetarians.

Its nutritional content shows it is a reasonably balanced meal which is high in energy with a good NSP content.

From the ingredients list it can be seen nuts have not been used so it is suitable for those with a nut allergy.

Convenience

It is ready for cooking and can be frozen.

It is quickly and conveniently prepared in the microwave.

It is good value for money.

Consumer protection

The weight of the product is displayed.

Also there is a use-by date, ingredients list and storage instructions.

If there is a complaint, the manufacturer's details are provided.

(12 marks)
London short course
higher 1998

Questions to test your design skills

Typical example: You have been approached by a large supermarket chain to develop a new soup. The product specification that you have been given is:

have a natural colour and taste, comparing well with home-made soup

appeals to a wide range of consumers

The new soup must

have a package which clearly conveys the home-made image

be suitable to serve for a special meal

Think about the new soup you would develop and explain how you will ensure that

(a) the specification is met.

(b) this specification will produce a soup which will sell. Give your reasons.

(a)   All natural ingredients must be used. A wide range of flavours must be available to suit different tastes. No additives or colouring are to be used. Tasting sessions and quality control are used to ensure successful results.

Unusual vegetables must be used and attractive garnishes or accompaniments suggested.

Packaging and advertising could convey a special image.

Eye-catching packaging with wording that shows a 'home-made' image. Must mention additives are not used and suggest a 'healthy' image.

Must have a range of flavours to cater for different consumers and different dietary needs.

It must be quick to prepare and at a reasonable price. Ideas must be trialled with a variety of consumers.

(b)   The soup may not appeal to consumers who are used to eating processed soups which do contain flavourings and colourings. These consumers may not enjoy a soup with a 'home-made' image so consumers may prefer something that looks more professional. The cost of a product suitable for a special occasion is often higher than everyday products so this will effect whether or not consumers buy it. A special occasion soup may not be popular if it is in a can or packet.

(6 + 4 marks)
(MEG short course
foundation 1998)

Nutritional values of food

Definitions

Nutrients provide part of the chemical make-up of food. Each nutrient has a specific function to perform within the body. After food has been digested the different nutrients are absorbed and used for the growth, repair or maintenance of body tissue.

Nutritional value of food is the number and type of nutrients provided by that food.

THE FUNCTION OF NUTRIENTS

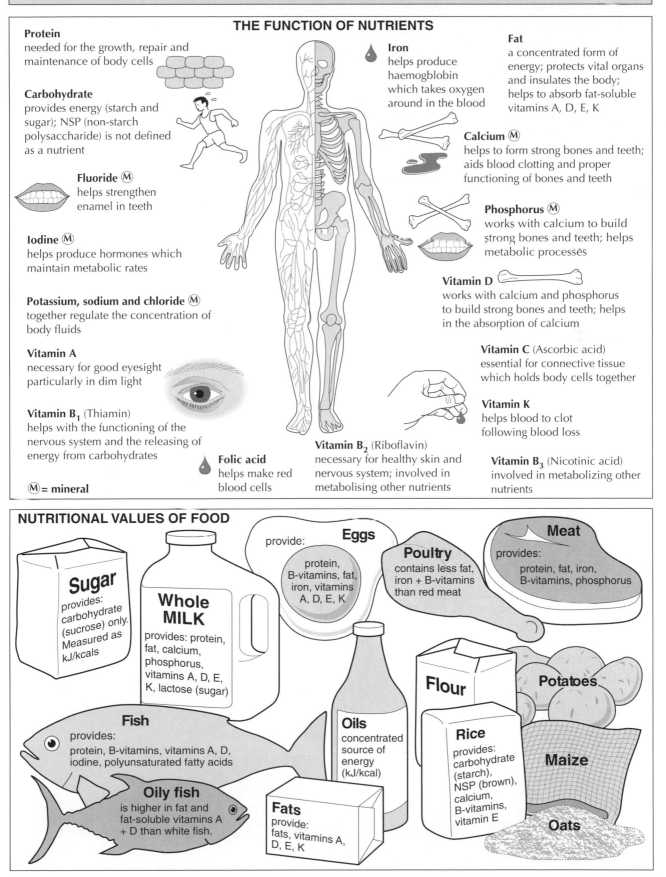

Protein
needed for the growth, repair and maintenance of body cells

Carbohydrate
provides energy (starch and sugar); NSP (non-starch polysaccharide) is not defined as a nutrient

Fluoride Ⓜ
helps strengthen enamel in teeth

Iodine Ⓜ
helps produce hormones which maintain metabolic rates

Potassium, sodium and chloride Ⓜ
together regulate the concentration of body fluids

Vitamin A
necessary for good eyesight particularly in dim light

Vitamin B$_1$ (Thiamin)
helps with the functioning of the nervous system and the releasing of energy from carbohydrates

Ⓜ = mineral

Iron
helps produce haemogblobin which takes oxygen around in the blood

Calcium Ⓜ
helps to form strong bones and teeth; aids blood clotting and proper functioning of bones and teeth

Phosphorus Ⓜ
works with calcium to build strong bones and teeth; helps metabolic processes

Vitamin D
works with calcium and phosphorus to build strong bones and teeth; helps in the absorption of calcium

Fat
a concentrated form of energy; protects vital organs and insulates the body; helps to absorb fat-soluble vitamins A, D, E, K

Vitamin C (Ascorbic acid)
essential for connective tissue which holds body cells together

Vitamin K
helps blood to clot following blood loss

Vitamin B$_3$ (Nicotinic acid)
involved in metabolizing other nutrients

Folic acid
helps make red blood cells

Vitamin B$_2$ (Riboflavin)
necessary for healthy skin and nervous system; involved in metabolising other nutrients

NUTRITIONAL VALUES OF FOOD

Sugar
provides: carbohydrate (sucrose) only. Measured as kJ/kcals

Whole MILK
provides: protein, fat, calcium, phosphorus, vitamins A, D, E, K, lactose (sugar)

Eggs
provide:
protein, B-vitamins, fat, iron, vitamins A, D, E, K

Poultry
contains less fat, iron + B-vitamins than red meat

Meat
provides:
protein, fat, iron, B-vitamins, phosphorus

Fish
provides:
protein, B-vitamins, vitamins A, D, iodine, polyunsaturated fatty acids

Oily fish
is higher in fat and fat-soluble vitamins A + D than white fish.

Fats
provide:
fats, vitamins A, D, E, K

Oils
concentrated source of energy (kJ/kcal)

Flour

Rice
provides:
carbohydrate (starch), NSP (brown), calcium, B-vitamins, vitamin E

Potatoes

Maize

Oats

Health aspects of food

Research linking diet and health and statistics showing a steady rise in certain diseases has led to dietary recommendations in the UK. Current dietary advice is based on government publications: *The Health of the Nation* (1992) and *Our Healthier Nation* (1998). Awareness of current health issues has led to an increased demand for 'healthy' food products.

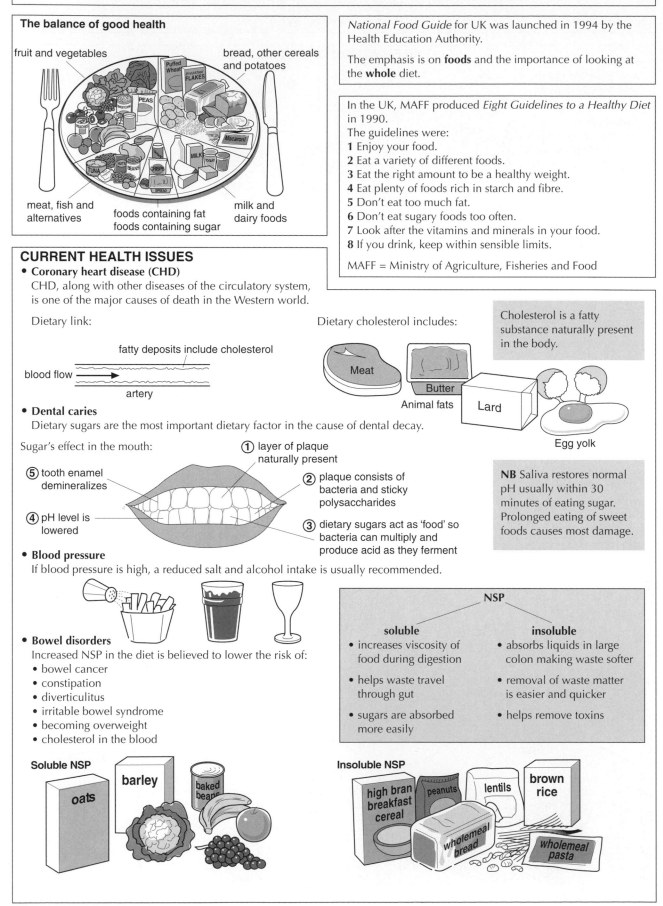

The balance of good health

fruit and vegetables

bread, other cereals and potatoes

meat, fish and alternatives

foods containing fat
foods containing sugar

milk and dairy foods

National Food Guide for UK was launched in 1994 by the Health Education Authority.

The emphasis is on **foods** and the importance of looking at the **whole** diet.

In the UK, MAFF produced *Eight Guidelines to a Healthy Diet* in 1990.
The guidelines were:
1 Enjoy your food.
2 Eat a variety of different foods.
3 Eat the right amount to be a healthy weight.
4 Eat plenty of foods rich in starch and fibre.
5 Don't eat too much fat.
6 Don't eat sugary foods too often.
7 Look after the vitamins and minerals in your food.
8 If you drink, keep within sensible limits.

MAFF = Ministry of Agriculture, Fisheries and Food

CURRENT HEALTH ISSUES

- **Coronary heart disease (CHD)**
  CHD, along with other diseases of the circulatory system, is one of the major causes of death in the Western world.

  Dietary link:

  fatty deposits include cholesterol

  blood flow

  artery

  Dietary cholesterol includes:

  Meat

  Butter
  Animal fats
  Lard

  Egg yolk

  Cholesterol is a fatty substance naturally present in the body.

- **Dental caries**
  Dietary sugars are the most important dietary factor in the cause of dental decay.

  Sugar's effect in the mouth:

  ① layer of plaque naturally present
  ② plaque consists of bacteria and sticky polysaccharides
  ③ dietary sugars act as 'food' so bacteria can multiply and produce acid as they ferment
  ④ pH level is lowered
  ⑤ tooth enamel demineralizes

  NB Saliva restores normal pH usually within 30 minutes of eating sugar. Prolonged eating of sweet foods causes most damage.

- **Blood pressure**
  If blood pressure is high, a reduced salt and alcohol intake is usually recommended.

- **Bowel disorders**
  Increased NSP in the diet is believed to lower the risk of:
  - bowel cancer
  - constipation
  - diverticulitus
  - irritable bowel syndrome
  - becoming overweight
  - cholesterol in the blood

  NSP

  soluble
  - increases viscosity of food during digestion
  - helps waste travel through gut
  - sugars are absorbed more easily

  insoluble
  - absorbs liquids in large colon making waste softer
  - removal of waste matter is easier and quicker
  - helps remove toxins

  Soluble NSP
  oats, barley, baked beans

  Insoluble NSP
  high bran breakfast cereal, peanuts, lentils, brown rice, wholemeal bread, wholemeal pasta

Physical and chemical properties of food : carbohydrates

Carbohydrates can be divided into **sugar**, **starch** and **non-starch polysaccharide (NSP)**.

Sugar is found naturally in a variety of foods. There are two forms:

Monosaccharides

fruits

vegetables

Disaccharides

Honey

sugar (sucrose)

milk (lactose)

shreddies malt products (maltose)

Properties

sweetens

TEA dissolves

improves the colour of baked products

caramelize

helps fat to incorporate more air in cake mixtures

acts as a preservative in jams and marmalades

prevents the development of gluten, making a softer cake or pastry

works with pectin to allow jams, marmalades, jellies to set when boiled with fruit

retains moisture in baked products and delays staling

strengthens the protein in whisked egg whites and helps to retain air

a sugar syrup prevents fruit from browning by excluding oxygen

provides a source of food for yeast to ferment

adds 'crunch' when sprinkled on foods

Starch is a polysaccharide naturally occurring in plants. Unlike sugar, it is not sweet and will not dissolve in cold water.

Vegetables

potatoes

turnips

parsnips

Cereal products

bread

flour

biscuits

pasta

cornflour

Cereals

wheat, oats, rice, maize

Properties

The role of starch in food production:

Sauces

Gelatinization:

starch granules

liquid

60 °C → starch granules swell as the temperature rises

85 °C thickened sauce

If allowed to cool the **gelatinized** liquid will **set** and form a **solid** sauce.

Structures

Starch, in the form of wheat flour, provides the structure of baked products

When bread is mixed with liquid in bread-making, the proteins (gliadin and glutenin) form another protein – gluten. Strong flour = high protein

Instant Dessert Flour can be treated to produce 'agglomerated' or instant flour which mixes easily with cold water

Wheat flour is commonly used in Britain

Flour forms the framework of a cake, for a soft crumbly texture, self-raising flour (low protein) is used

The gluten in flour sets on baking.

When products containing starch in the form of flour are baked, the starch **gelatinizes**, the proteins **coagulate** and the product **sets** in its risen position.

Non-starch polysaccharide (NSP) occurs with starch in foods of plant origin. It comes from the cell walls of plants and consists of pectins, celluloses, hemicelluloses and lignin.

Further details are provided on page 2.

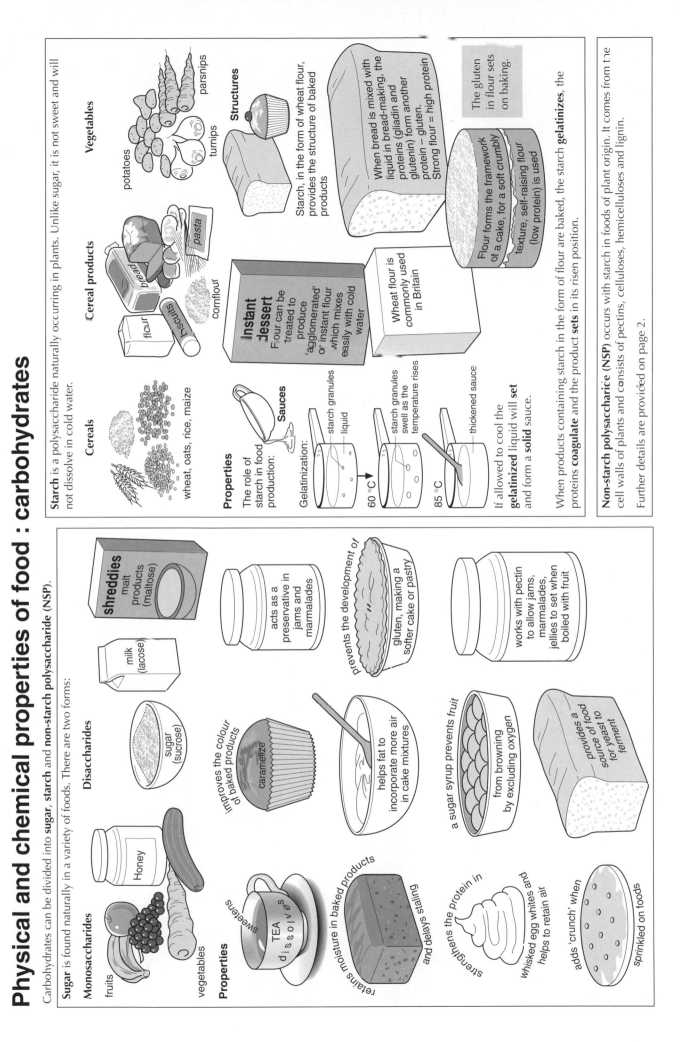

Physical and chemical properties of food : protein and fat

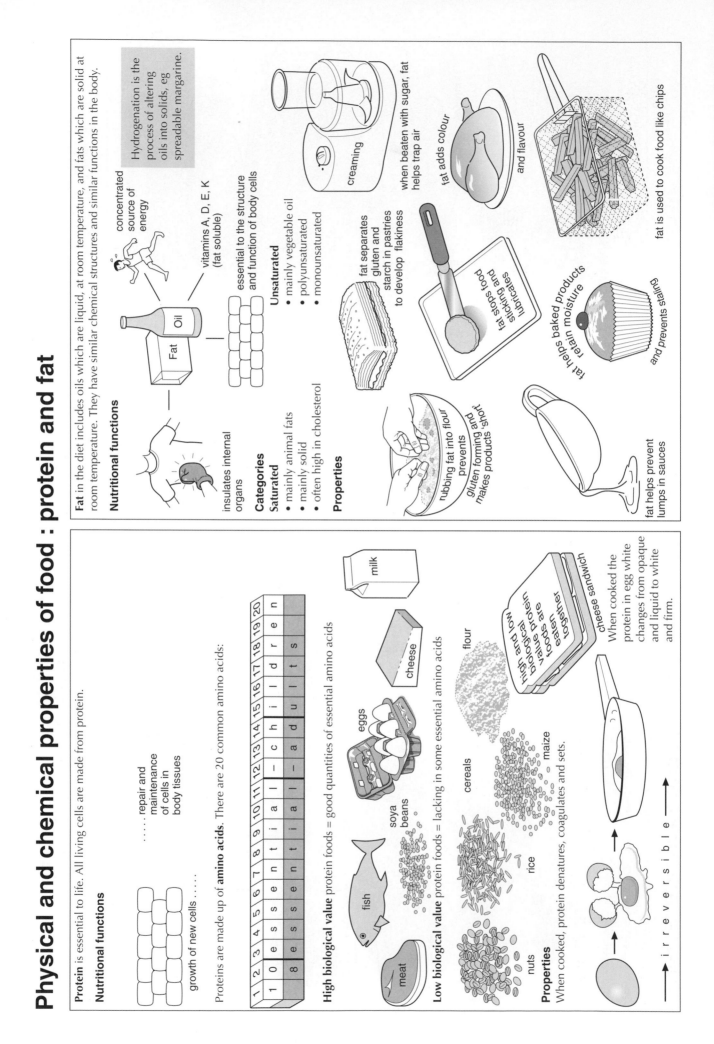

Physical and chemical properties of food (Protein)

Protein is essential to life. All living cells are made from protein.

Nutritional functions

..... repair and maintenance of cells in body tissues

growth of new cells .....

Proteins are made up of **amino acids**. There are 20 common amino acids:

1	2	3	4	5	6	7	8	9	10	11	12	13	14	15	16	17	18	19	20
10		e	s	s	e	n	t	i	a	l	–	c	h	i	l	d	r	e	n
8		e	s	s	e	n	t	i	a	l	–	a	d	u	l	t	s		

High biological value protein foods = good quantities of essential amino acids

soya beans, fish, meat, eggs, milk

Low biological value protein foods = lacking in some essential amino acids

nuts, rice, cereals, maize, flour, cheese

high and low biological value protein foods are eaten together — cheese sandwich

Properties

When cooked, protein denatures, coagulates and sets.

i r r e v e r s i b l e

When cooked the protein in egg white changes from opaque and liquid to white and firm.

Fat

Fat in the diet includes oils which are liquid, at room temperature, and fats which are solid at room temperature. They have similar chemical structures and similar functions in the body.

Nutritional functions

concentrated source of energy

vitamins A, D, E, K (fat soluble)

essential to the structure and function of body cells

insulates internal organs

Hydrogenation is the process of altering oils into solids, eg spreadable margarine.

Categories

Saturated
- mainly animal fats
- mainly solid
- often high in cholesterol

Unsaturated
- mainly vegetable oil
- polyunsaturated
- monounsaturated

Properties

Creaming — when beaten with sugar, fat helps trap air

fat adds colour and flavour

fat is used to cook food like chips

fat separates gluten and starch in pastries to develop flakiness

fat stops food sticking and lubricates

fat helps baked products retain moisture and prevents staling

rubbing fat into flour prevents gluten forming and makes products 'short'

fat helps prevent lumps in sauces

Sensory characteristics of food

Definition

A sensory **characteristic** of a food is a 'feature' that involves our senses. For example, red in colour or sour in taste.
All foods are a mixture of different sensory characteristics. The characteristics of an ingredient will affect the resulting food product.

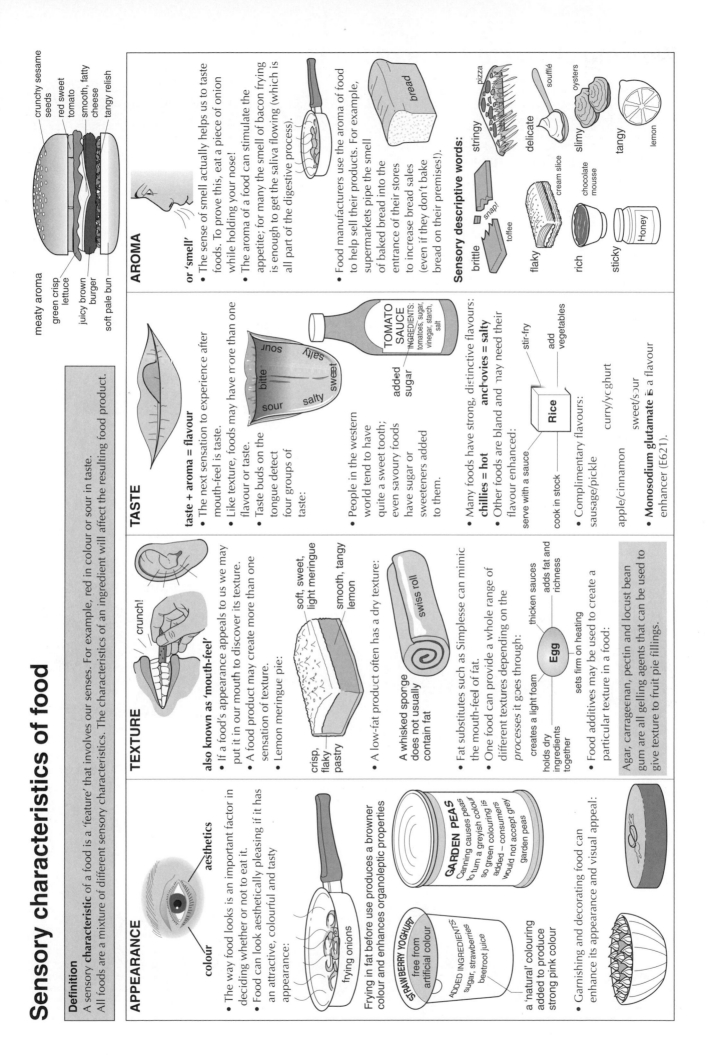

- crunchy sesame seeds
- red sweet tomato
- smooth, fatty cheese
- tangy relish
- meaty aroma
- green crisp lettuce
- juicy brown burger
- soft pale bun

APPEARANCE

colour   aesthetics

- The way food looks is an important factor in deciding whether or not to eat it.
- Food can look aesthetically pleasing if it has an attractive, colourful and tasty appearance:

frying onions

Frying in fat before use produces a browner colour and enhances organoleptic properties

GARDEN PEAS
Canning causes peas to turn a greyish colour so green colouring is added – consumers would not accept grey garden peas

STRAWBERRY YOGHURT
free from artificial colour
ADDED INGREDIENTS
Sugar, strawberries
beetroot juice

a 'natural' colouring added to produce strong pink colour

- Garnishing and decorating food can enhance its appearance and visual appeal:

TEXTURE

crunch!

also known as 'mouth-feel'

- If a food's appearance appeals to us we may put it in our mouth to discover its texture.
- A food product may create more than one sensation of texture.
- Lemon meringue pie:

  crisp, flaky pastry

  soft, sweet, light meringue

  smooth, tangy lemon

- A low-fat product often has a dry texture:

  swiss roll

  A whisked sponge does not usually contain fat

- Fat substitutes such as Simplesse can mimic the mouth-feel of fat.
- One food can provide a whole range of different textures depending on the *processes* it goes through:

  Egg
  - creates a light foam
  - holds dry ingredients together
  - thicken sauces
  - adds fat and richness
  - sets firm on heating

- Food additives may be used to create a particular texture in a food:

Agar, carrageenan, pectin and locust bean gum are all gelling agents that can be used to give texture to fruit pie fillings.

TASTE

taste + aroma = flavour

- The next sensation to experience after mouth-feel is taste.
- Like texture, foods may have more than one flavour or taste.
- Taste buds on the tongue detect four groups of taste:

  sour
  bitter
  sour
  salty
  salty
  sweet

- People in the western world tend to have quite a sweet tooth; even savoury foods have sugar or sweeteners added to them.

  TOMATO SAUCE
  INGREDIENTS: tomatoes, sugar, vinegar, starch, salt

  added sugar

- Many foods have strong, distinctive flavours:

  chillies = hot   **anchovies = salty**

- Other foods are bland and may need their flavour enhanced:

  serve with a sauce — stir-fry
  cook in stock — **Rice** — add vegetables

- Complimentary flavours:

  sausage/pickle
  apple/cinnamon
  curry/yoghurt
  sweet/sour

- **Monosodium glutamate** is a flavour enhancer (E621).

AROMA

or 'smell'

- The sense of smell actually helps us to taste foods. To prove this, eat a piece of onion while holding your nose!
- The aroma of a food can stimulate the appetite; for many the smell of bacon frying is enough to get the saliva flowing (which is all part of the digestive process).

- Food manufacturers use the aroma of food to help sell their products. For example, supermarkets pipe the smell of baked bread into the entrance of their stores to increase bread sales (even if they don't bake bread on their premises!).

  bread

Sensory descriptive words:

- pizza
- stringy
- soufflé
- oysters
- delicate
- lemon
- tangy
- brittle
- snap!
- cream slice
- chocolate mousse
- slimy
- toffee
- flaky
- rich
- Honey
- sticky

Working characteristics of food

Definition
The working characteristics of food are the factors which influence the outcome of a product: appearance, flavour, colour, volume, shape, palatability, finish.

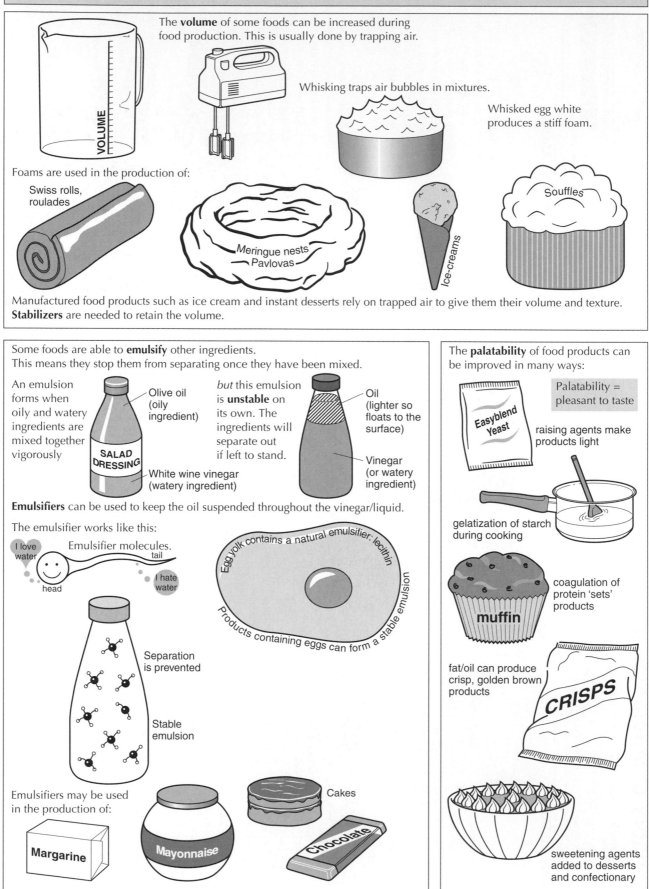

The **volume** of some foods can be increased during food production. This is usually done by trapping air.

Whisking traps air bubbles in mixtures.

Whisked egg white produces a stiff foam.

Foams are used in the production of:

Swiss rolls, roulades

Meringue nests
Pavlovas

Ice-creams

Souffles

Manufactured food products such as ice cream and instant desserts rely on trapped air to give them their volume and texture. **Stabilizers** are needed to retain the volume.

Some foods are able to **emulsify** other ingredients. This means they stop them from separating once they have been mixed.

An emulsion forms when oily and watery ingredients are mixed together vigorously

Olive oil (oily ingredient)

White wine vinegar (watery ingredient)

SALAD DRESSING

but this emulsion is **unstable** on its own. The ingredients will separate out if left to stand.

Oil (lighter so floats to the surface)

Vinegar (or watery ingredient)

Emulsifiers can be used to keep the oil suspended throughout the vinegar/liquid.

The emulsifier works like this:

I love water
head
Emulsifier molecules.
tail
I hate water

Egg yolk contains a natural emulsifier: lecithin

Products containing eggs can form a stable emulsion

Separation is prevented

Stable emulsion

Emulsifiers may be used in the production of:

Margarine

Mayonnaise

Cakes

Chocolate

The **palatability** of food products can be improved in many ways:

Palatability = pleasant to taste

Easyblend Yeast

raising agents make products light

gelatization of starch during cooking

muffin

coagulation of protein 'sets' products

fat/oil can produce crisp, golden brown products

CRISPS

sweetening agents added to desserts and confectionary

Primary to secondary sources of food : wheat

Definition

Primary sources of food are the raw materials (eg wheat) which will become the ingredients or secondary sources of food (eg flour) after being processed.

A **wheatgrain:**
- member of cereals family
- other members include rice, barley, rye, maize, oats
- grown for their grains (seeds)

Endosperm
Bran
Germ

Hair
Crease
Plumule
Radicle

Three main parts:
1 bran coating (approx. 15%)
2 endosperm (approx. 82.5%)
3 germ (embryo) (approx. 2.5%)

PRIMARY FOOD: WHEAT

Cleaning

beard removed

dry cleaning and sieving

dry cleaning

rotary grading

metals extracted

spiral separation

washing

Conditioning

distribution of moisture

drying

ready for milling

Milling - stage 1

endosperm graded

bran coat removed

Milling - stage 2

rolling and sieving

granular flour

Mixing

Sterilizing and packing

sterilizing

automatic packing

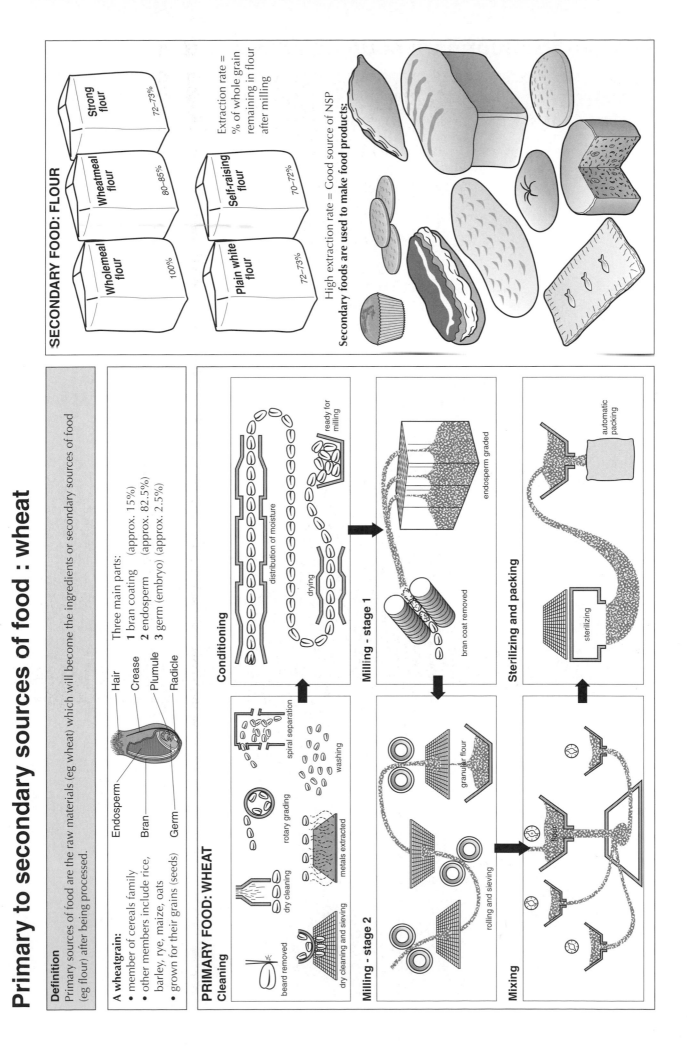

SECONDARY FOOD: FLOUR

Strong flour 72–73%

Wholemeal flour 100%

Wheatmeal flour 80–85%

Plain white flour 72–73%

Self-raising flour 70–72%

Extraction rate = % of whole grain remaining in flour after milling

High extraction rate = Good source of NSP

Secondary foods are used to make food products:

Primary to secondary sources of food : soya

Definition
A versatile protein food is produced from the soya bean.

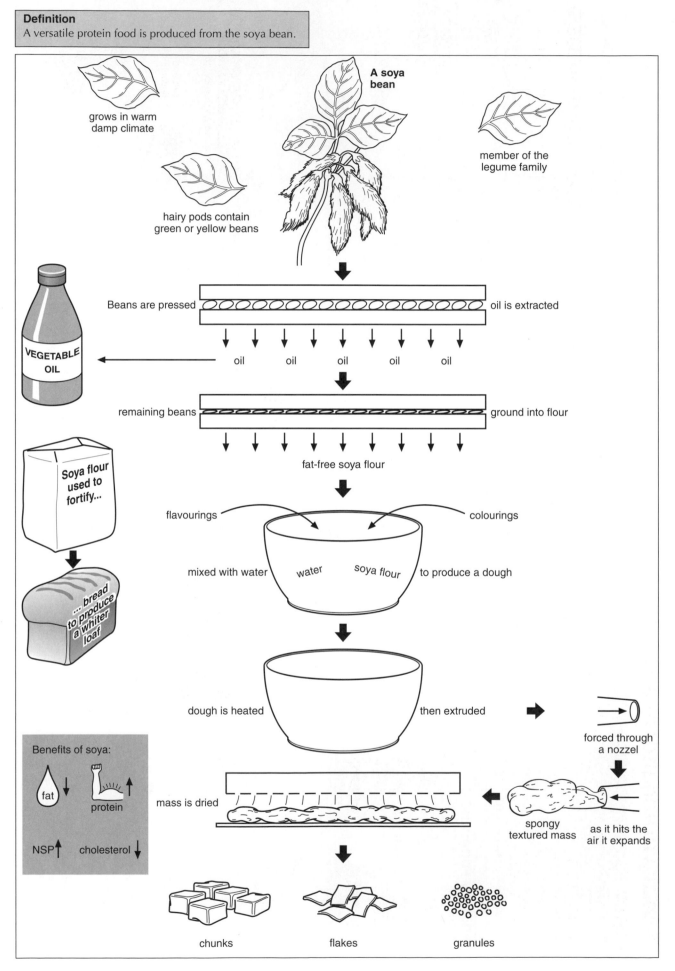

A soya bean

grows in warm damp climate

hairy pods contain green or yellow beans

member of the legume family

Beans are pressed — oil is extracted

VEGETABLE OIL

oil   oil   oil   oil   oil

remaining beans — ground into flour

fat-free soya flour

Soya flour used to fortify...

...bread to produce a whiter loaf

flavourings — colourings

mixed with water   water   soya flour   to produce a dough

dough is heated   then extruded

forced through a nozzle

Benefits of soya:

fat ↓   protein ↑

NSP ↑   cholesterol ↓

mass is dried

spongy textured mass   as it hits the air it expands

chunks   flakes   granules

Manufacturing processes

Definition: Manufacturing processes are the methods used to produce food products.
The choice of manufacturing process usually depends on the quantity of food products being produced.

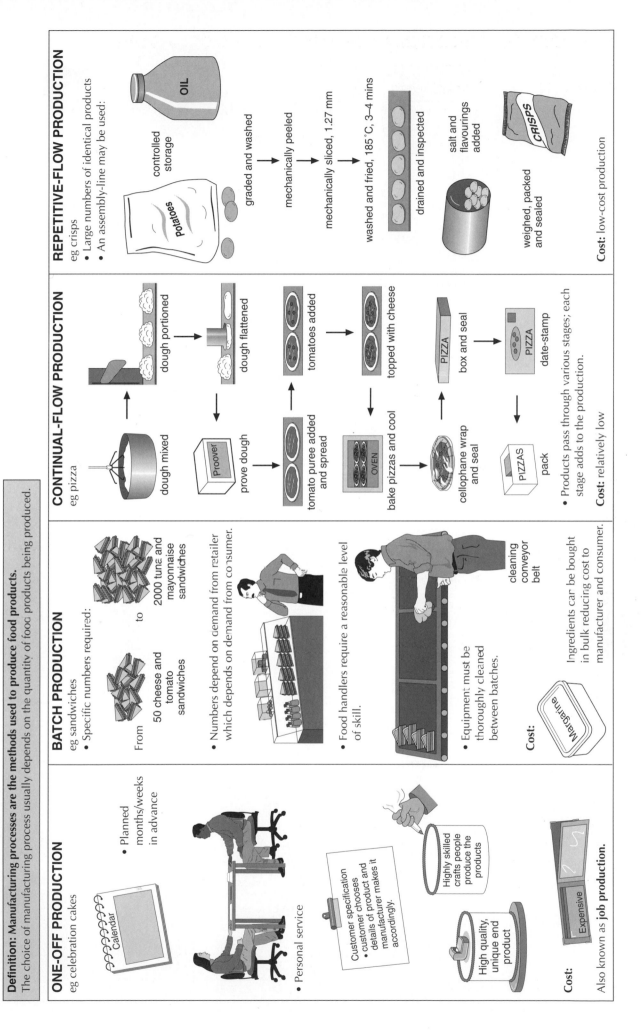

ONE-OFF PRODUCTION
eg celebration cakes

- Planned months/weeks in advance

Calendar

- Personal service

Customer specification
• customer chooses details of product and manufacturer makes it accordingly.

Highly skilled crafts people produce the products

High quality, unique end product

Cost: Expensive

Also known as **job production.**

BATCH PRODUCTION
eg sandwiches

- Specific numbers required:

From   50 cheese and tomato sandwiches   to   2000 tuna and mayonnaise sandwiches

- Numbers depend on demand from retailer which depends on demand from consumer.

- Food handlers require a reasonable level of skill.

- Equipment must be thoroughly cleaned between batches.

cleaning conveyor belt

Margarine

Cost: Ingredients can be bought in bulk reducing cost to manufacturer and consumer.

CONTINUAL-FLOW PRODUCTION
eg pizza

dough mixed → dough portioned → dough flattened

Proover — prove dough

tomato puree added and spread → tomatoes added → topped with cheese

OVEN — bake pizzas and cool

cellophane wrap and seal

PIZZA — box and seal → PIZZA — date-stamp

PIZZAS — pack

- Products pass through various stages; each stage adds to the production.

Cost: relatively low

REPETITIVE-FLOW PRODUCTION
eg crisps

- Large numbers of identical products
- An assembly-line may be used:

OIL

potatoes — controlled storage

graded and washed

mechanically peeled

mechanically sliced, 1.27 mm

washed and fried, 185°C, 3–4 mins

drained and inspected

salt and flavourings added

CRISPS

weighed, packed and sealed

Cost: low-cost production

Heat transference and heat processes

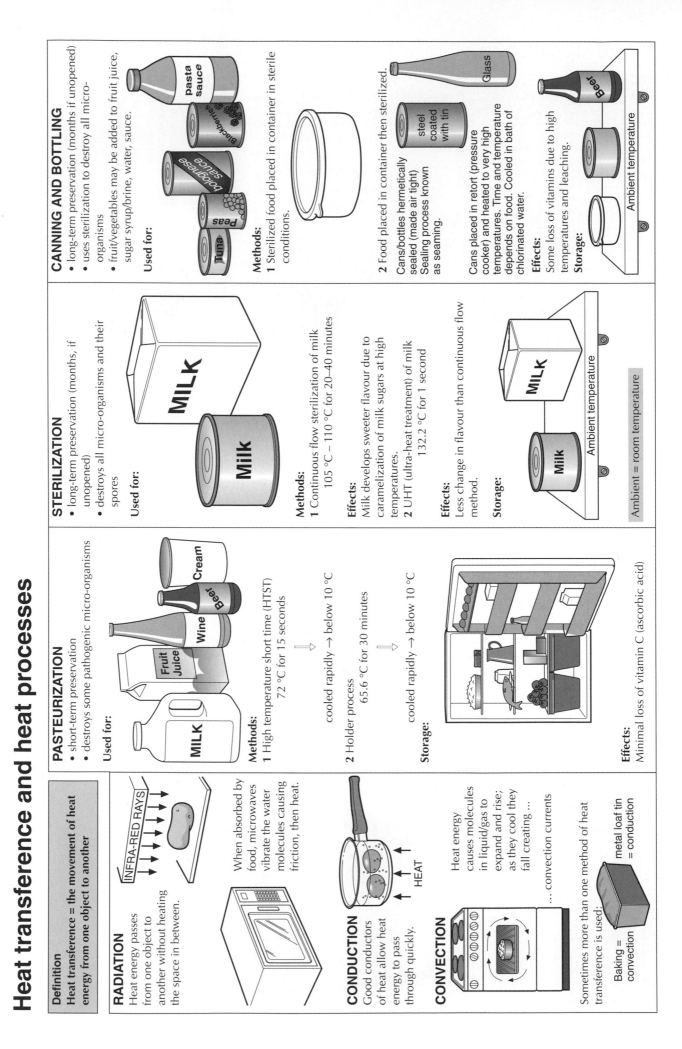

Definition
Heat transference = the movement of heat energy from one object to another

RADIATION
Heat energy passes from one object to another without heating the space in between.

INFRA-RED RAYS

When absorbed by food, microwaves vibrate the water molecules causing friction, then heat.

CONDUCTION
Good conductors of heat allow heat energy to pass through quickly.

HEAT

CONVECTION
Heat energy causes molecules in liquid/gas to expand and rise; as they cool they fall creating … … convection currents

Sometimes more than one method of heat transference is used:

metal loaf tin = conduction

Baking = convection

PASTEURIZATION
- short-term preservation
- destroys some pathogenic micro-organisms

Used for:
MILK   Fruit Juice   Wine   Beer   Cream

Methods:
1 High temperature short time (HTST)
72 °C for 15 seconds

⇨ cooled rapidly → below 10 °C

2 Holder process
65.6 °C for 30 minutes

⇨ cooled rapidly → below 10 °C

Storage:

Effects:
Minimal loss of vitamin C (ascorbic acid)

STERILIZATION
- long-term preservation (months, if unopened)
- destroys all micro-organisms and their spores

Used for:
MILK   Milk

Methods:
1 Continuous flow sterilization of milk
105 °C – 110 °C for 20–40 minutes

Effects:
Milk develops sweeter flavour due to caramelization of milk sugars at high temperatures.

2 UHT (ultra-heat treatment) of milk
132.2 °C for 1 second

Effects:
Less change in flavour than continuous flow method.

Storage:
MILK   Milk

Ambient temperature

Ambient = room temperature

CANNING AND BOTTLING
- long-term preservation (months if unopened)
- uses sterilization to destroy all micro-organisms
- fruit/vegetables may be added to fruit juice, sugar syrup/brine, water, sauce.

Used for:
pasta sauce   Blackberries   bolognese sauce   Peas   Tuna

Methods:
1 Sterilized food placed in container in sterile conditions.

2 Food placed in container then sterilized.
Cans/bottles hermetically sealed (made air tight)
Sealing process known as seaming.

Glass   steel coated with tin

Cans placed in retort (pressure cooker) and heated to very high temperatures. Time and temperature depends on food. Cooled in bath of chlorinated water.

Effects:
Some loss of vitamins due to high temperatures and leaching.

Storage:
Beer

Ambient temperature

Food spoilage

Definition
Food spoilage refers to food that appears unpalatable such as sour milk or a brown apple. It is a natural process, speeded up by enzymes present in food.

ENZYMIC BROWNING/OXIDATION

Food spoilage is not harmful but makes food look unpleasant.

The action of enzymes can be **reduced** by:

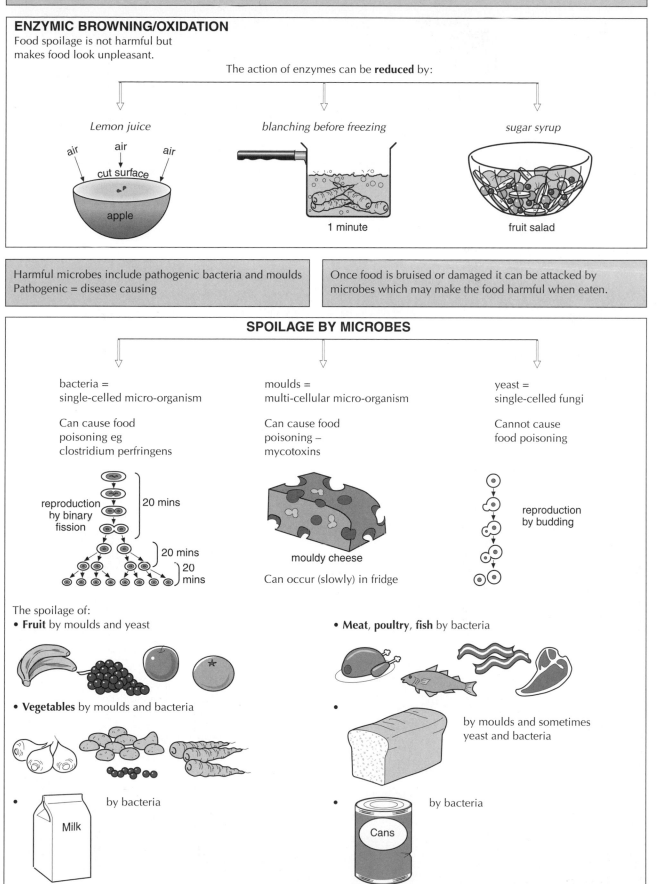

Lemon juice

air · air · air
cut surface
apple

blanching before freezing

1 minute

sugar syrup

fruit salad

Harmful microbes include pathogenic bacteria and moulds
Pathogenic = disease causing

Once food is bruised or damaged it can be attacked by microbes which may make the food harmful when eaten.

SPOILAGE BY MICROBES

bacteria =
single-celled micro-organism

Can cause food poisoning eg clostridium perfringens

reproduction by binary fission — 20 mins
20 mins
20 mins

moulds =
multi-cellular micro-organism

Can cause food poisoning – mycotoxins

mouldy cheese

Can occur (slowly) in fridge

yeast =
single-celled fungi

Cannot cause food poisoning

reproduction by budding

The spoilage of:
• **Fruit** by moulds and yeast

• **Vegetables** by moulds and bacteria

• by bacteria

Milk

• **Meat**, **poultry**, **fish** by bacteria

• by moulds and sometimes yeast and bacteria

• by bacteria

Cans

Food contamination and food poisoning

Definitions
Food contamination is the presence of any unacceptable matter in food (micro-organisms, foreign bodies, poisons).
Food poisoning is the result of consuming foods contaminated with moulds, viruses, bacteria, or their toxins, or chemicals, or naturally toxic plants and animals.

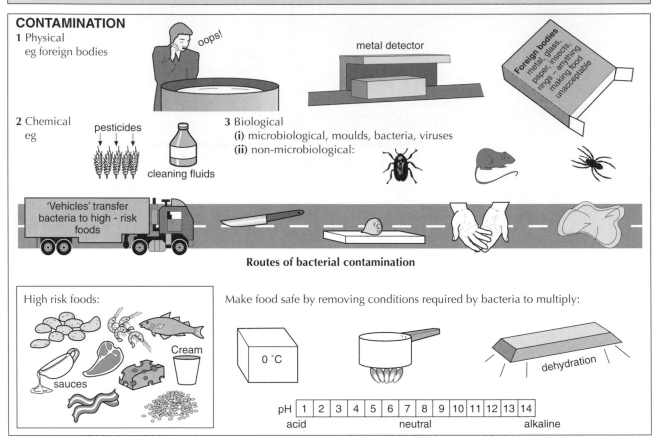

CONTAMINATION

1 Physical
eg foreign bodies

oops!

metal detector

Foreign bodies
metal, glass, paper, insects, rings – anything making food unacceptable

2 Chemical
eg

pesticides

cleaning fluids

3 Biological
(i) microbiological, moulds, bacteria, viruses
(ii) non-microbiological:

'Vehicles' transfer bacteria to high - risk foods

Routes of bacterial contamination

High risk foods:

Cream

sauces

Make food safe by removing conditions required by bacteria to multiply:

0 °C

dehydration

pH	1	2	3	4	5	6	7	8	9	10	11	12	13	14

acid   neutral   alkaline

FOOD POISONING

Name	Source	Symptoms	Control measures
Salmonella		fever vomiting diarrhoea stomach pains	Heat ↑ 75 °C Store ↓ 5 °C Defrost thoroughly
Campylobacter		fever diarrhoea	Cook meat thoroughly
Listeria monocytogenes	salad Pate Brie	flu-like symptoms meningitus septicaemia	Store ↓ 5 °C If young, old, pregnant or ill, avoid soft cheeses, pates
Escherichia (E.coli)		vomiting bloody diarrhoea kidney disease	Cook meat thoroughly. Avoid contamination
Staphylococcus aureus	hands nose mouth	severe stomach pain, vomiting, lowering of body temperature	High standards of personal hygiene for food handlers

Food and low temperatures

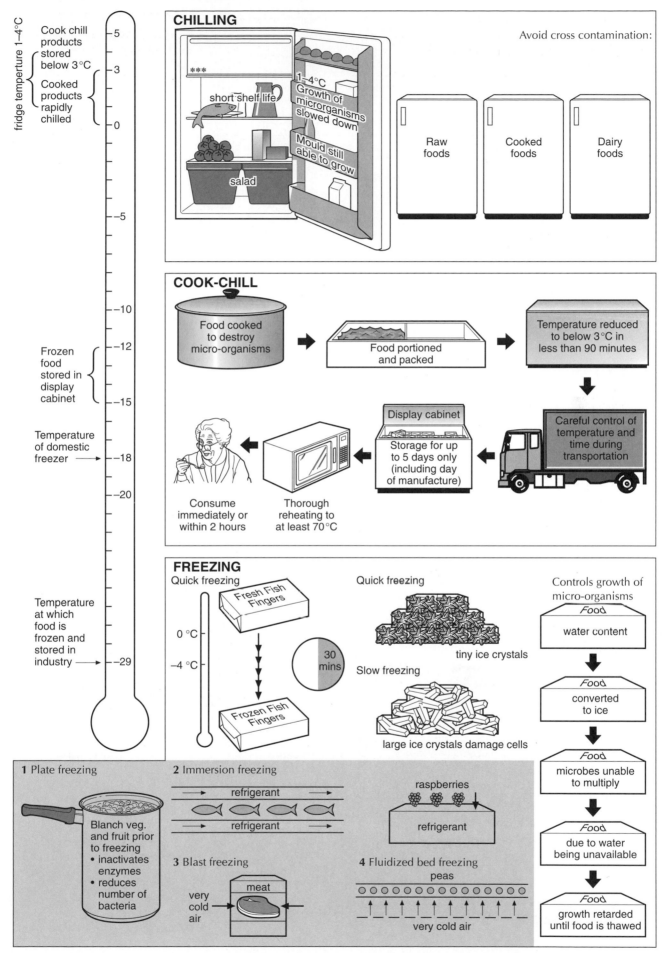

Methods of gathering information

In order to design and develop a new food product it is necessary to gather and interpret information.

What information is needed?

Inspiration for new ideas

What consumers think of existing products

What consumers think of new products

Evidence that there is a market for a new product

How can the information be gathered?
1. Gathering information as inspiration ...

Browse through cookery books

Existing products

What is already for sale?

ADVERT

What has just been launched onto the market?

cookery books

Trade magazines

The Grocer

Look at magazines

THE GOOD FOOD SHOW

Visit food shows and exhibitions

Who's buying what?

Try out new products

Eat out

2. Gathering information from consumers often involves **market research**.
Market research is the process of gathering information about a product or market. It is used to help food companies make decisions. It is usually carried out at some stage during new product development.

Quantitative research is carried out on a large scale; attempts to measure a response; may be used to compare products; occurs at advanced stage of development.

I wouldn't pay more than £3.99

Why is that?

I always buy chicken twice a week

Qualitative research reports consumer **views** but does not quantify them; useful when new product and packaging are at concept stage.

NEW PRODUCT

3. Gathering information by means of survey or questionnaire is known as **primary research**.

Secondary research involves obtaining information from databases, reports, surveys already carried out, books and other documents.

Designing a survey or questionnaire

- gather only relevant information by planning questions carefully
- check the order of the questions is logical
- take account of the target consumer group
- carry out a test run to ensure questions can be understood

Types of questions

- closed
  eg Do you own a microwave: Yes/No
- open
  eg What do you like about the product's appearance?
- structured
  eg In your opinion which statement best describes the flavour
  a) rich and creamy
  b) dull and bland

Analyzing information

If the information gathered is going to be of value it needs to be analyzed and interpreted.

Qualitative research involves finding out what consumers think; this may not be easy to interpret, eg
Focus Group results: What do you think of the product's appearance?

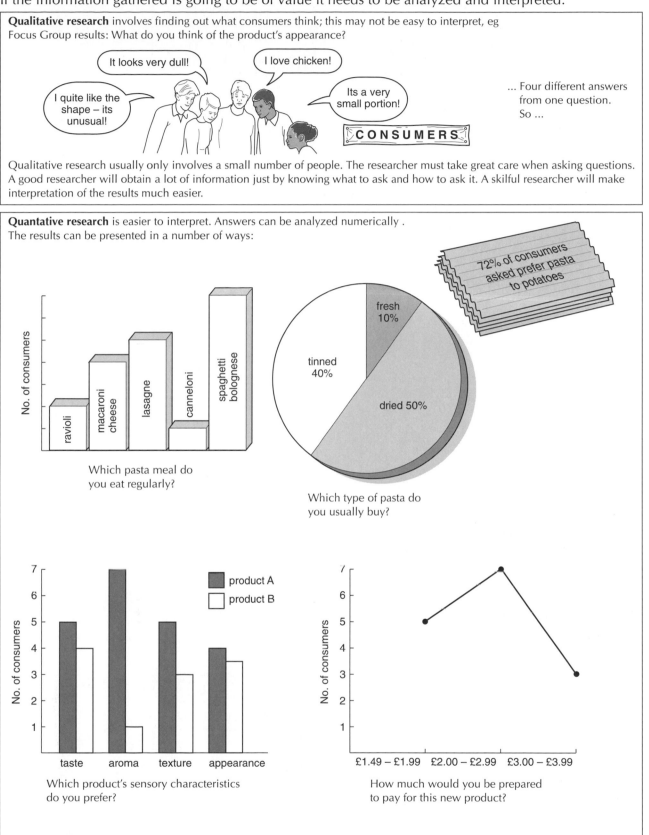

It looks very dull!

I love chicken!

I quite like the shape – its unusual!

Its a very small portion!

CONSUMERS

... Four different answers from one question.
So ...

Qualitative research usually only involves a small number of people. The researcher must take great care when asking questions. A good researcher will obtain a lot of information just by knowing what to ask and how to ask it. A skilful researcher will make interpretation of the results much easier.

Quantative research is easier to interpret. Answers can be analyzed numerically .
The results can be presented in a number of ways:

72% of consumers asked prefer pasta to potatoes

No. of consumers

ravioli | macaroni cheese | lasagne | canneloni | spaghetti bolognese

Which pasta meal do you eat regularly?

fresh 10%

tinned 40%

dried 50%

Which type of pasta do you usually buy?

No. of consumers

product A
product B

taste | aroma | texture | appearance

Which product's sensory characteristics do you prefer?

No. of consumers

£1.49 – £1.99   £2.00 – £2.99   £3.00 – £3.99

How much would you be prepared to pay for this new product?

ANALYZING THE RESULTS

After gathering and interpreting the information, the results must be analyzed. If questions have been carefully chosen and worded, the results should prove useful during Product Development. A new product idea can be modified or altered to meet consumer needs. The results may indicate there is not a market for the new product after all and a great deal of money will not be wasted on its production.

Comparing a student-made food product with a commercial equivalent

HOW DO MANUFACTURED PRODUCTS COMPARE WITH THOSE MADE IN FOOD TECHNOLOGY?

Student-made chocolate ice-cream	Brand X chocolate ice-cream	Brand Y chocolate ice-cream
Ingredients: 300 ml milk 100 g plain chocolate 3 egg yolks 50 g sugar 300 ml whipping cream 50 g chocolate chips	**Ingredients:** fresh cream skimmed milk Belgian chocolate sugar chocolate fudge pieces egg yolk coco natural flavouring: vanilla	**Ingredients:** reconstituted skimmed milk sugar butter chocolate chocolate flavour ripple (contains stabilizers: E410, E412, E415; flavouring: citric acid) coco powder egg yolk emulsifier: E471 stabilizers: E412, E402

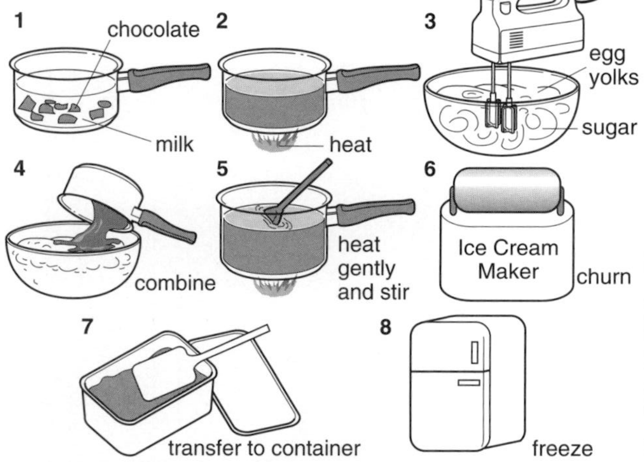

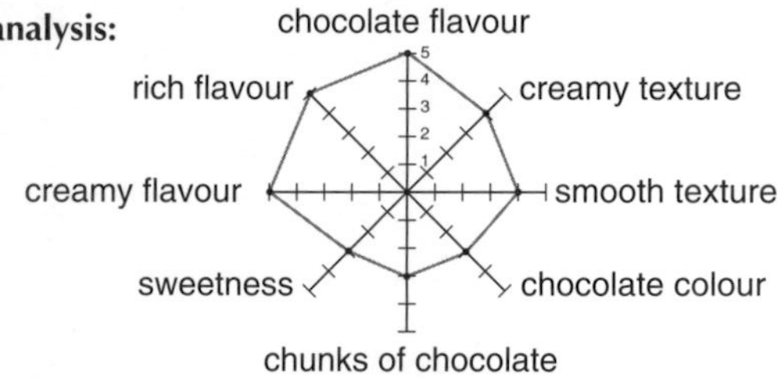

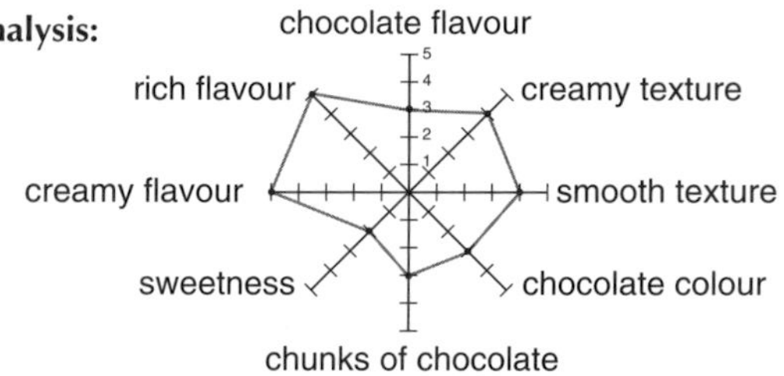

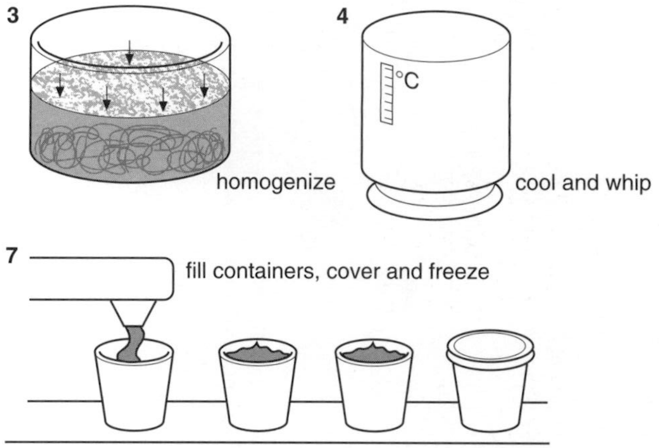

Production (Student-made):

1 chocolate / milk — heat
2 heat
3 egg yolks / sugar
4 combine
5 heat gently and stir
6 Ice Cream Maker — churn
7 transfer to container
8 freeze

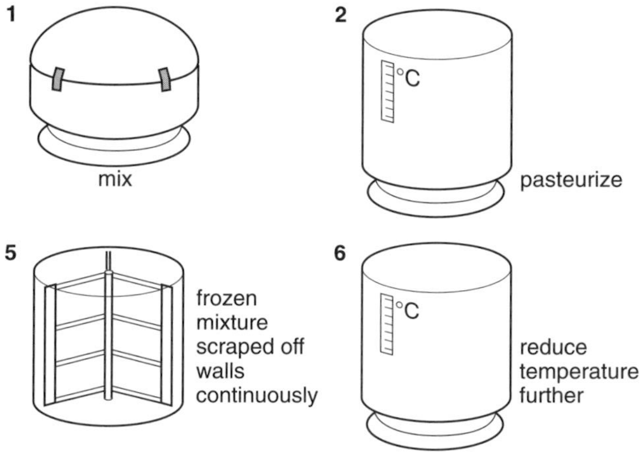

Production (Brand X):

1 mix
2 °C pasteurize
3 homogenize
5 frozen mixture scraped off walls continuously
6 °C reduce temperature further

Production (Brand Y):

3 homogenize
4 °C cool and whip
7 fill containers, cover and freeze

Shelf-life: 1 week | **Shelf-life:** 1 year | **Shelf-life:** 9 months

Sensory analysis:

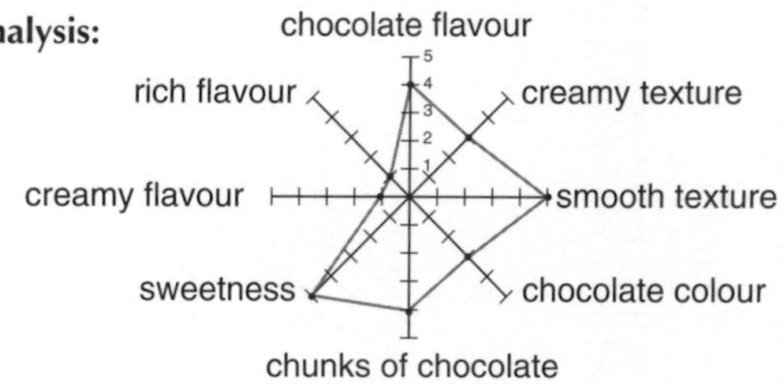

(radar charts showing: chocolate flavour, creamy texture, smooth texture, chocolate colour, chunks of chocolate, sweetness, creamy flavour, rich flavour)

Developing a new food product based on a domestic recipe requires **investigation** and **careful adaptation**.

Comparing and disassembling existing food products

When developing new food products it is important to look at similar products already on the market.

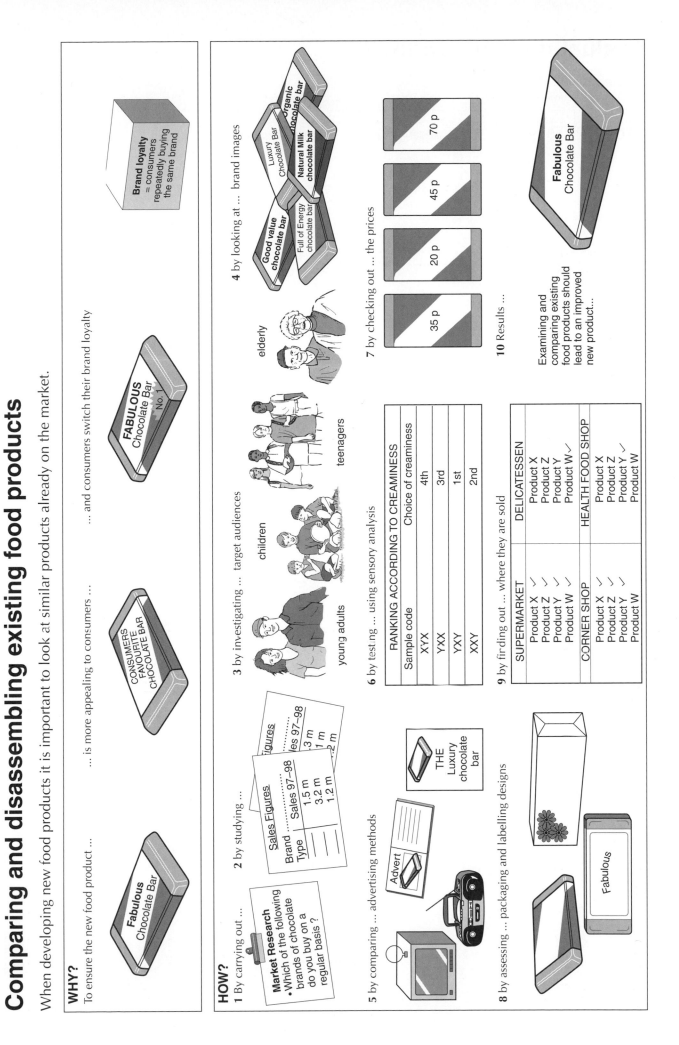

WHY?

To ensure the new food product ... ... is more appealing to consumers ... ... and consumers switch their brand loyalty

Brand loyalty = consumers repeatedly buying the same brand

Fabulous Chocolate Bar

CONSUMERS FAVOURITE CHOCOLATE BAR

FABULOUS Chocolate Bar No. 1

HOW?

1 By carrying out ...

Market Research
• Which of the following brands of chocolate do you buy on a regular basis?

2 by studying ...

Sales Figures

Brand	Sales 97–98
	1.5 m
	3.2 m
	1.2 m

Sales Figures

Type	Sales 97–98
	.3 m
	1 m
	.2 m

3 by investigating ... target audiences

young adults   children   teenagers   elderly

4 by looking at ... brand images

Luxury Chocolate Bar   Organic chocolate bar   Good value chocolate bar   Natural Milk chocolate bar   Full of Energy chocolate bar

5 by comparing ... advertising methods

Advert   THE Luxury chocolate bar

6 by testing ... using sensory analysis

RANKING ACCORDING TO CREAMINESS

Sample code	Choice of creaminess
X Y X	4th
Y X X	3rd
Y X Y	1st
X X Y	2nd

7 by checking out ... the prices

35 p   20 p   45 p   70 p

8 by assessing ... packaging and labelling designs

Fabulous

9 by finding out ... where they are sold

SUPERMARKET		DELICATESSEN	
Product X	✓✓	Product X	
Product Z	✓✓	Product Z	
Product Y	✓✓	Product Y	✓
Product W	✓	Product W	✓
CORNER SHOP		HEALTH FOOD SHOP	
Product X	✓✓	Product X	
Product Z	✓✓	Product Z	
Product Y	✓✓	Product Y	✓
Product W		Product W	

10 Results ...

Examining and comparing existing food products should lead to an improved new product...

Fabulous Chocolate Bar

Evaluating food products

WHEN TO EVALUATE?

designing

Throughout designing process

making   testing

developing

HOW TO EVALUATE?
1. Choose criteria.
2. Decide on suitable evaluation method for each criteria.

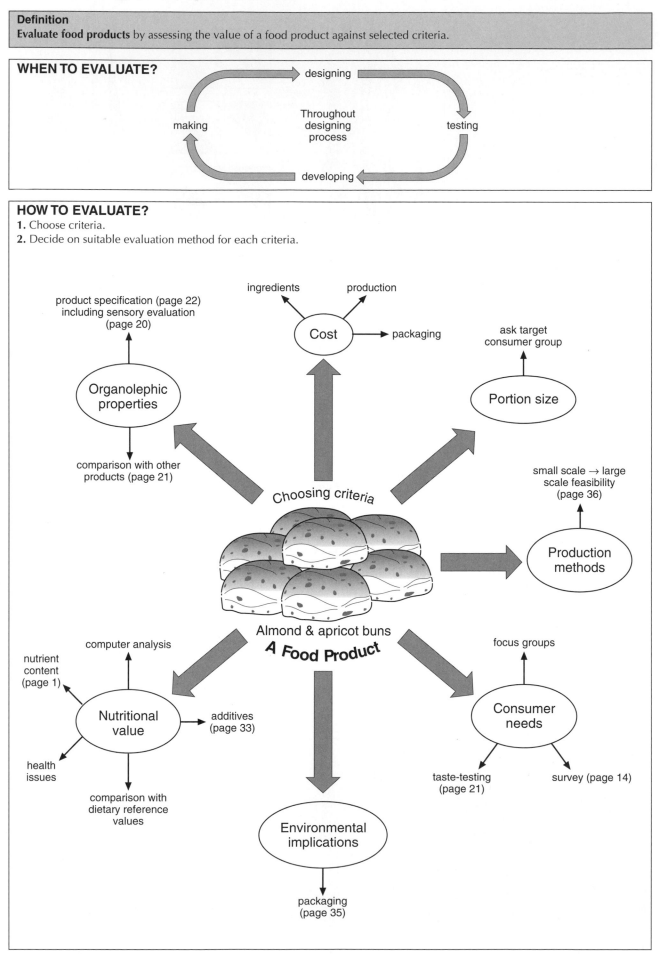

product specification (page 22)
including sensory evaluation
(page 20)

ingredients   production

Cost → packaging

ask target
consumer group

Organolephic
properties

Portion size

comparison with other
products (page 21)

small scale → large
scale feasibility
(page 36)

Choosing criteria

Production
methods

nutrient
content
(page 1)

computer analysis

Nutritional
value → additives
(page 33)

Almond & apricot buns
A Food Product

focus groups

Consumer
needs

health
issues

comparison with
dietary reference
values

taste-testing
(page 21)

survey (page 14)

Environmental
implications

packaging
(page 35)

Modifying recipes

Definition
To **modify a recipe** is to adapt or change one or more ingredient(s) in type or quantity, or alter process(es) of the production.

WHY MODIFY?

In order to improve a product prototype after testing and evaluating it

In order to satisfy a particular design specification eg meeting current dietary guidelines

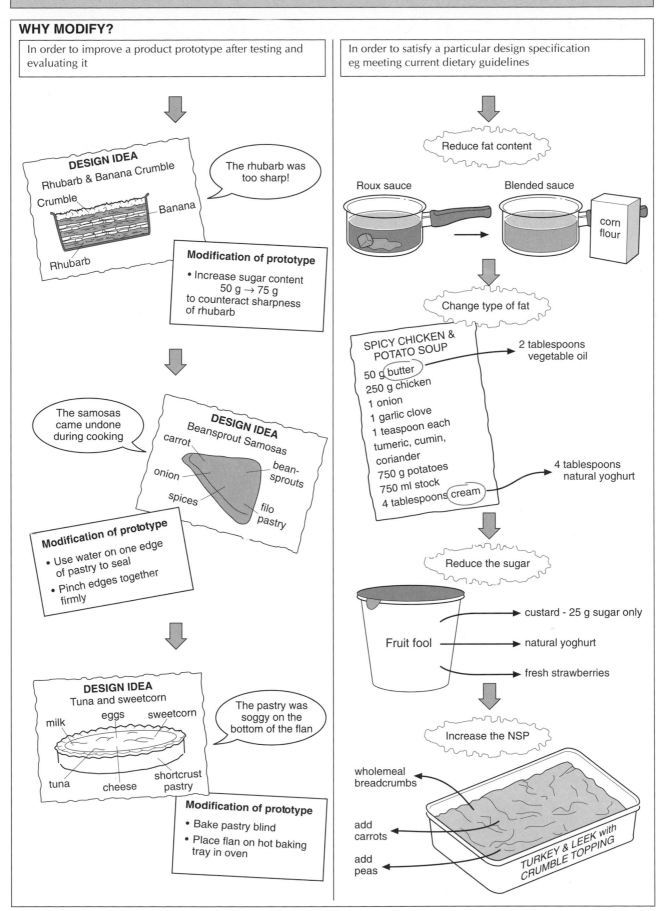

DESIGN IDEA
Rhubarb & Banana Crumble
Crumble
Banana
Rhubarb

The rhubarb was too sharp!

Modification of prototype
• Increase sugar content
50 g → 75 g
to counteract sharpness of rhubarb

The samosas came undone during cooking

DESIGN IDEA
Beansprout Samosas
carrot
bean-sprouts
onion
spices
filo pastry

Modification of prototype
• Use water on one edge of pastry to seal
• Pinch edges together firmly

DESIGN IDEA
Tuna and sweetcorn
milk
eggs
sweetcorn
tuna
cheese
shortcrust pastry

The pastry was soggy on the bottom of the flan

Modification of prototype
• Bake pastry blind
• Place flan on hot baking tray in oven

Reduce fat content

Roux sauce
Blended sauce
corn flour

Change type of fat

SPICY CHICKEN & POTATO SOUP
50 g butter → 2 tablespoons vegetable oil
250 g chicken
1 onion
1 garlic clove
1 teaspoon each tumeric, cumin, coriander
750 g potatoes
750 ml stock
4 tablespoons cream → 4 tablespoons natural yoghurt

Reduce the sugar

Fruit fool
custard - 25 g sugar only
natural yoghurt
fresh strawberries

Increase the NSP

wholemeal breadcrumbs
add carrots
add peas
TURKEY & LEEK with CRUMBLE TOPPING

Sensory analysis 1

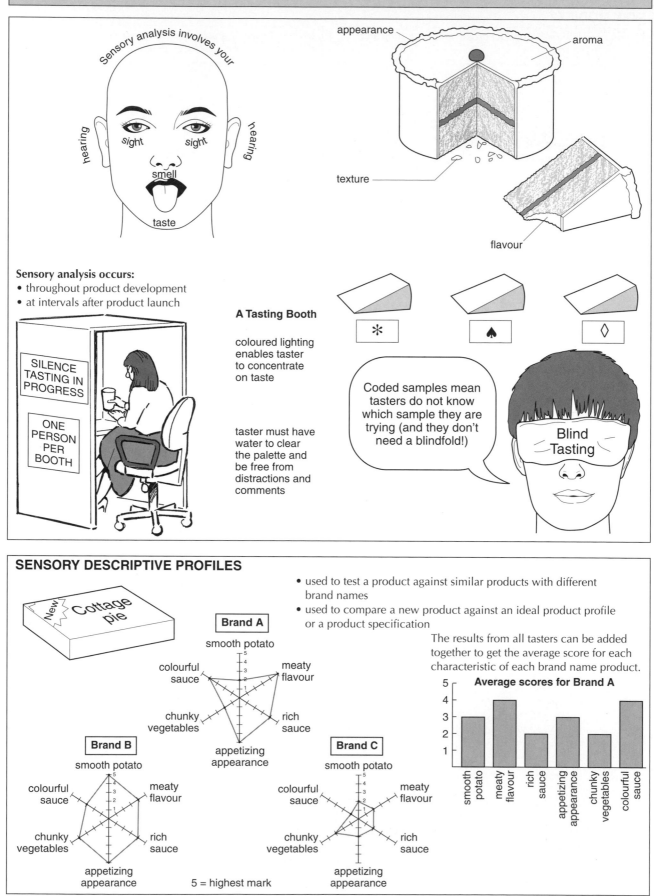

Sensory analysis involves your

sight   sight

hearing   hearing

smell

taste

appearance

aroma

texture

flavour

Sensory analysis occurs:
- throughout product development
- at intervals after product launch

SILENCE TASTING IN PROGRESS

ONE PERSON PER BOOTH

A Tasting Booth

coloured lighting enables taster to concentrate on taste

taster must have water to clear the palette and be free from distractions and comments

Coded samples mean tasters do not know which sample they are trying (and they don't need a blindfold!)

Blind Tasting

SENSORY DESCRIPTIVE PROFILES

New Cottage Pie

- used to test a product against similar products with different brand names
- used to compare a new product against an ideal product profile or a product specification

The results from all tasters can be added together to get the average score for each characteristic of each brand name product.

Brand A

smooth potato

colourful sauce   meaty flavour

chunky vegetables   rich sauce

appetizing appearance

Brand B

smooth potato

colourful sauce   meaty flavour

chunky vegetables   rich sauce

appetizing appearance

5 = highest mark

Brand C

smooth potato

colourful sauce   meaty flavour

chunky vegetables   rich sauce

appetizing appearance

Average scores for Brand A

smooth potato, meaty flavour, rich sauce, appetizing appearance, chunky vegetables, colourful sauce

Sensory analysis 2

△ TRIANGLE TEST

The team of tasters are each given a set of 3 coded samples. The samples are presented randomly. The number of correct answers are added up. The reason for the difference is not required.

Can you detect a difference? eg

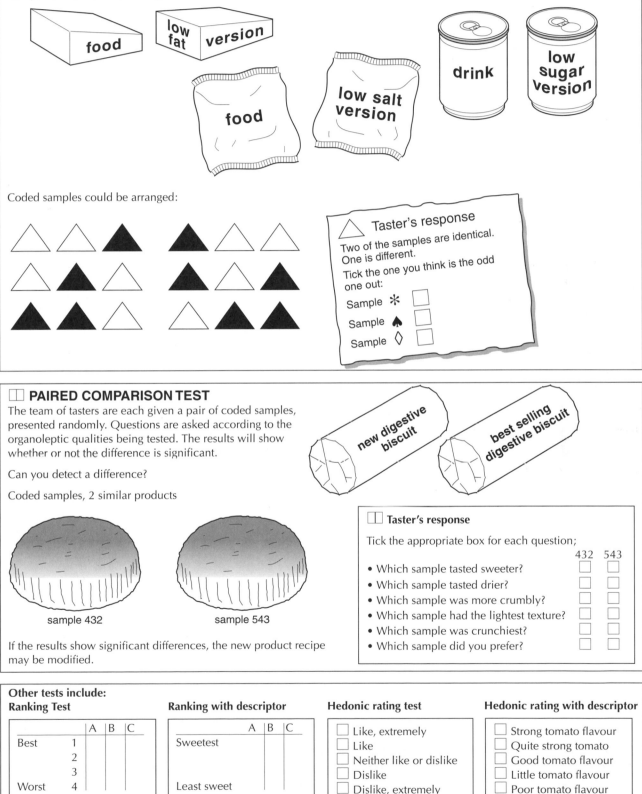

Coded samples could be arranged:

△ Taster's response

Two of the samples are identical.
One is different.
Tick the one you think is the odd one out:

Sample ✳ ☐

Sample ♠ ☐

Sample ◊ ☐

☐ PAIRED COMPARISON TEST

The team of tasters are each given a pair of coded samples, presented randomly. Questions are asked according to the organoleptic qualities being tested. The results will show whether or not the difference is significant.

new digestive biscuit *best selling digestive biscuit*

Can you detect a difference?

Coded samples, 2 similar products

sample 432   sample 543

If the results show significant differences, the new product recipe may be modified.

☐ Taster's response

Tick the appropriate box for each question;

	432	543
• Which sample tasted sweeter?	☐	☐
• Which sample tasted drier?	☐	☐
• Which sample was more crumbly?	☐	☐
• Which sample had the lightest texture?	☐	☐
• Which sample was crunchiest?	☐	☐
• Which sample did you prefer?	☐	☐

Other tests include:

Ranking Test

		A	B	C
Best	1			
	2			
	3			
Worst	4			

Ranking with descriptor

	A	B	C
Sweetest			
Least sweet			

Hedonic rating test

☐ Like, extremely
☐ Like
☐ Neither like or dislike
☐ Dislike
☐ Dislike, extremely

Hedonic rating with descriptor

☐ Strong tomato flavour
☐ Quite strong tomato
☐ Good tomato flavour
☐ Little tomato flavour
☐ Poor tomato flavour

Specifications

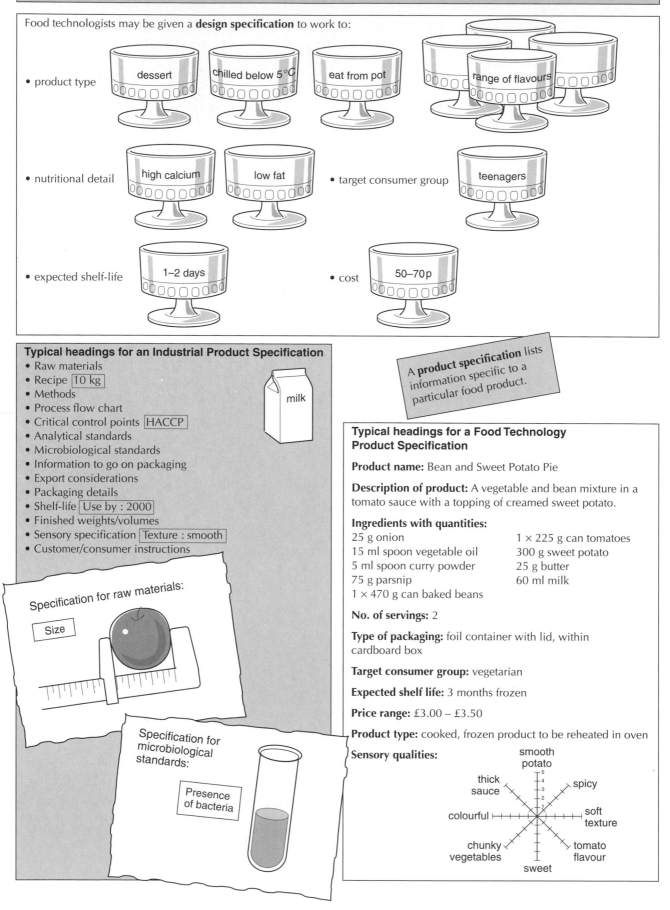

Food technologists may be given a **design specification** to work to:

- product type
  - dessert
  - chilled below 5°C
  - eat from pot
  - range of flavours

- nutritional detail
  - high calcium
  - low fat

- target consumer group
  - teenagers

- expected shelf-life
  - 1–2 days

- cost
  - 50–70p

Typical headings for an Industrial Product Specification

- Raw materials
- Recipe 10 kg
- Methods
- Process flow chart
- Critical control points HACCP
- Analytical standards
- Microbiological standards
- Information to go on packaging
- Export considerations
- Packaging details
- Shelf-life Use by : 2000
- Finished weights/volumes
- Sensory specification Texture : smooth
- Customer/consumer instructions

milk

A **product specification** lists information specific to a particular food product.

Specification for raw materials:

Size

Specification for microbiological standards:

Presence of bacteria

Typical headings for a Food Technology Product Specification

Product name: Bean and Sweet Potato Pie

Description of product: A vegetable and bean mixture in a tomato sauce with a topping of creamed sweet potato.

Ingredients with quantities:

25 g onion	1 × 225 g can tomatoes
15 ml spoon vegetable oil	300 g sweet potato
5 ml spoon curry powder	25 g butter
75 g parsnip	60 ml milk
1 × 470 g can baked beans	

No. of servings: 2

Type of packaging: foil container with lid, within cardboard box

Target consumer group: vegetarian

Expected shelf life: 3 months frozen

Price range: £3.00 – £3.50

Product type: cooked, frozen product to be reheated in oven

Sensory qualities:

smooth potato
thick sauce
spicy
colourful
soft texture
chunky vegetables
tomato flavour
sweet

Testing products against the specification

Throughout the development of a new food product, evaluation is essential. One method of evaluating is to test against the product specification.

After making a new product, check the specification details:

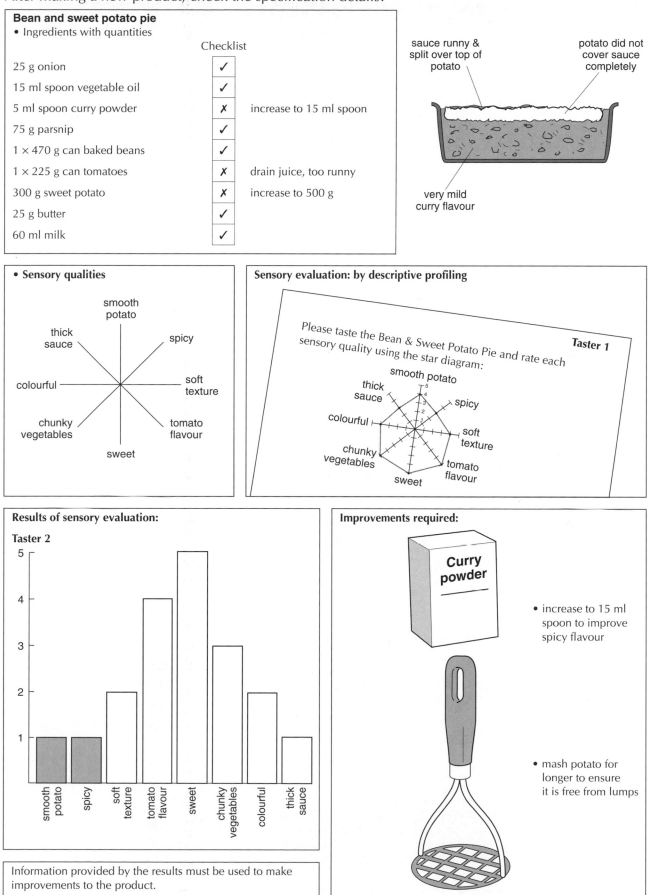

Bean and sweet potato pie
• Ingredients with quantities

		Checklist	
25 g onion		✓	
15 ml spoon vegetable oil		✓	
5 ml spoon curry powder		✗	increase to 15 ml spoon
75 g parsnip		✓	
1 × 470 g can baked beans		✓	
1 × 225 g can tomatoes		✗	drain juice, too runny
300 g sweet potato		✗	increase to 500 g
25 g butter		✓	
60 ml milk		✓	

sauce runny & split over top of potato

potato did not cover sauce completely

very mild curry flavour

• **Sensory qualities**

smooth potato

thick sauce

spicy

colourful

soft texture

chunky vegetables

tomato flavour

sweet

Sensory evaluation: by descriptive profiling

Please taste the Bean & Sweet Potato Pie and rate each sensory quality using the star diagram: **Taster 1**

smooth potato

thick sauce

spicy

colourful

soft texture

chunky vegetables

tomato flavour

sweet

Results of sensory evaluation:

Taster 2

Improvements required:

Curry powder

• increase to 15 ml spoon to improve spicy flavour

• mash potato for longer to ensure it is free from lumps

Information provided by the results must be used to make improvements to the product.

Target consumer groups and their needs 1

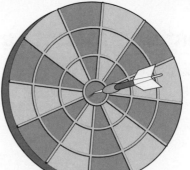

Definition
Target consumer groups are the consumers for whom a product is designed.

A new product will only succeed if there are consumers who need, want and will repeatedly buy that product.

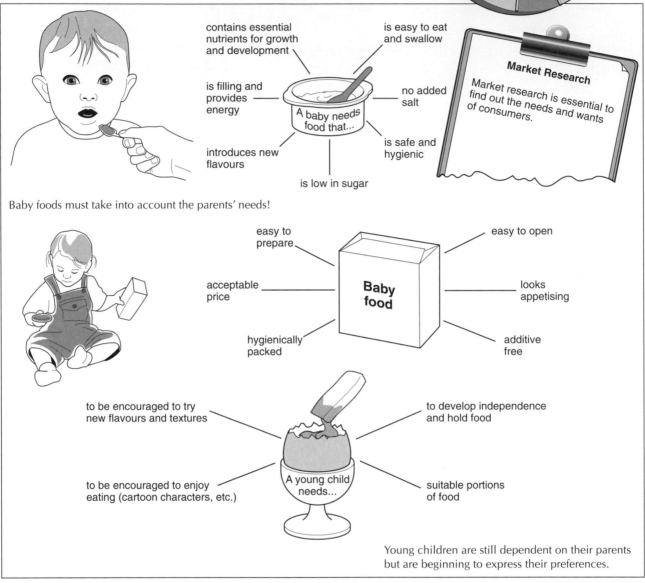

contains essential nutrients for growth and development

is filling and provides energy

introduces new flavours

A baby needs food that...

is easy to eat and swallow

no added salt

is safe and hygienic

is low in sugar

Market Research
Market research is essential to find out the needs and wants of consumers.

Baby foods must take into account the parents' needs!

easy to prepare

acceptable price

hygienically packed

Baby food

easy to open

looks appetising

additive free

to be encouraged to try new flavours and textures

to be encouraged to enjoy eating (cartoon characters, etc.)

A young child needs...

to develop independence and hold food

suitable portions of food

Young children are still dependent on their parents but are beginning to express their preferences.

Teenagers are very much influenced by the outside world and not just their parents.

Teenagers need a balanced diet that fulfils their specific energy requirements.

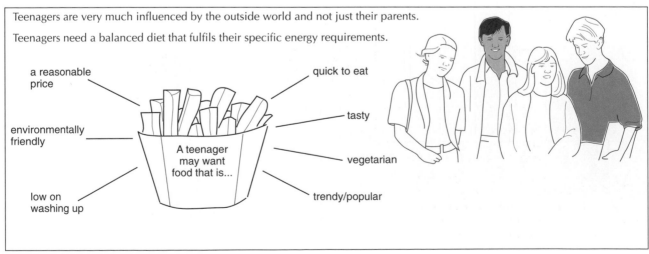

a reasonable price

environmentally friendly

low on washing up

A teenager may want food that is...

quick to eat

tasty

vegetarian

trendy/popular

Target consumer groups and their needs 2

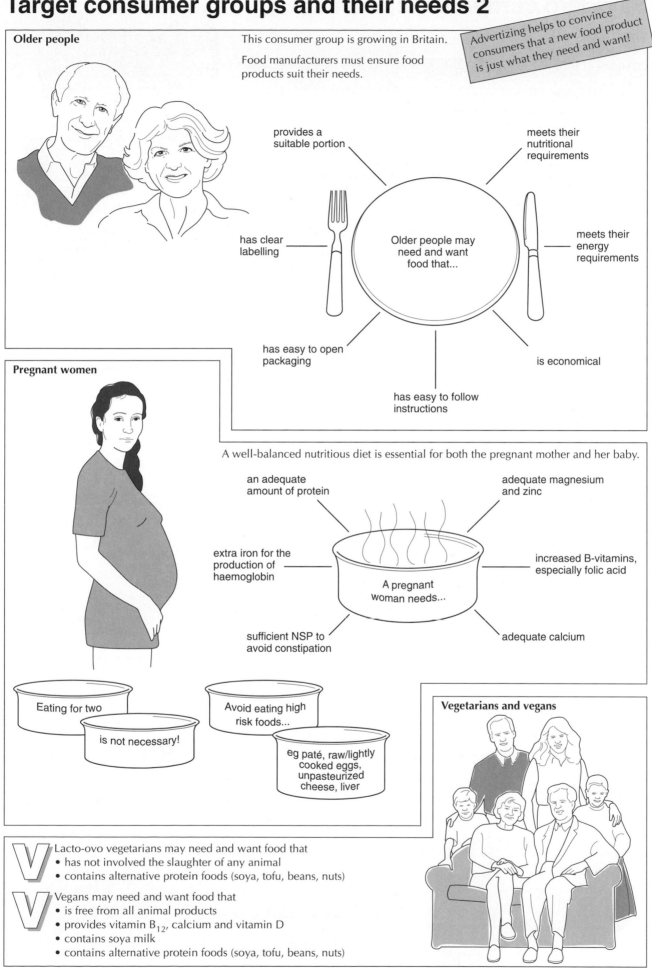

Older people

This consumer group is growing in Britain.

Food manufacturers must ensure food products suit their needs.

Advertizing helps to convince consumers that a new food product is just what they need and want!

- provides a suitable portion
- meets their nutritional requirements
- has clear labelling
- Older people may need and want food that...
- meets their energy requirements
- has easy to open packaging
- is economical
- has easy to follow instructions

Pregnant women

A well-balanced nutritious diet is essential for both the pregnant mother and her baby.

- an adequate amount of protein
- adequate magnesium and zinc
- extra iron for the production of haemoglobin
- A pregnant woman needs...
- increased B-vitamins, especially folic acid
- sufficient NSP to avoid constipation
- adequate calcium

Eating for two

is not necessary!

Avoid eating high risk foods...

eg paté, raw/lightly cooked eggs, unpasteurized cheese, liver

Vegetarians and vegans

V Lacto-ovo vegetarians may need and want food that
- has not involved the slaughter of any animal
- contains alternative protein foods (soya, tofu, beans, nuts)

VV Vegans may need and want food that
- is free from all animal products
- provides vitamin B_{12}, calcium and vitamin D
- contains soya milk
- contains alternative protein foods (soya, tofu, beans, nuts)

Standard components

ADVANTAGES of using standard components:

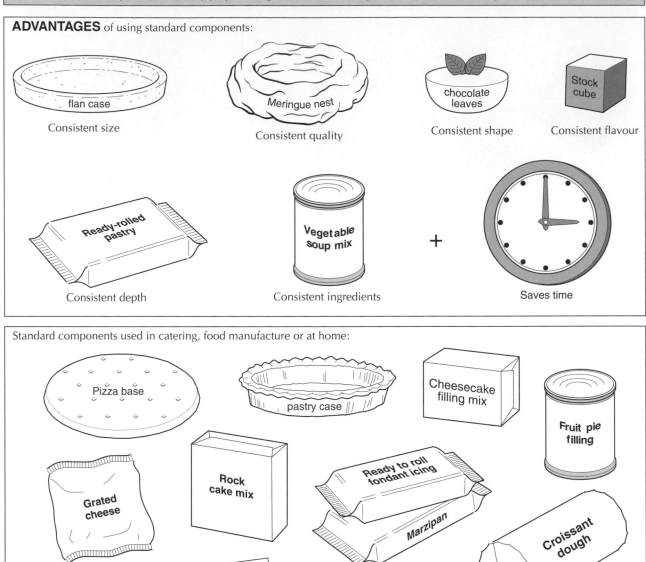

flan case
Consistent size

Meringue nest
Consistent quality

chocolate leaves
Consistent shape

Stock cube
Consistent flavour

Ready-rolled pastry
Consistent depth

Vegetable soup mix
Consistent ingredients

+
Saves time

Standard components used in catering, food manufacture or at home:

Pizza base

pastry case

Cheesecake filling mix

Fruit pie filling

Grated cheese

Rock cake mix

Ready to roll fondant icing

Marzipan

Croissant dough

Using standard components ensures consistent and reliable results.

Potential problem:

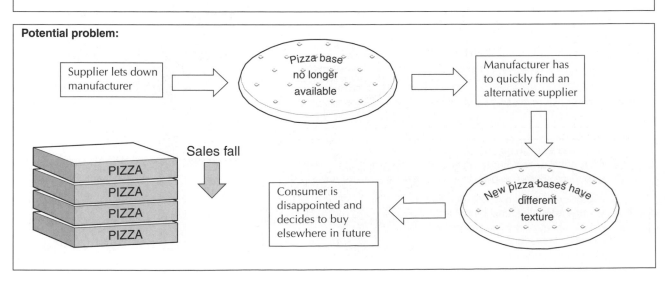

Supplier lets down manufacturer

Pizza base no longer available

Manufacturer has to quickly find an alternative supplier

New pizza bases have different texture

Consumer is disappointed and decides to buy elsewhere in future

Sales fall

PIZZA
PIZZA
PIZZA
PIZZA

Quality assurance

Definition
Quality assurance is a term used by food manufacturers to guarantee products are of a particular standard.

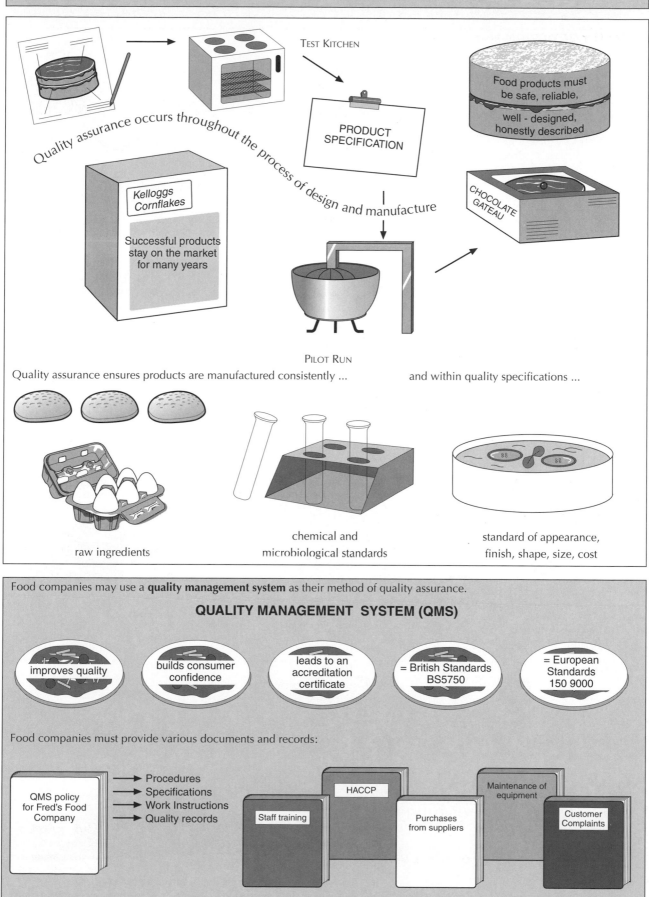

TEST KITCHEN

PRODUCT SPECIFICATION

Quality assurance occurs throughout the process of design and manufacture

Food products must be safe, reliable, well-designed, honestly described

Kelloggs Cornflakes

Successful products stay on the market for many years

CHOCOLATE GATEAU

PILOT RUN

Quality assurance ensures products are manufactured consistently ...

and within quality specifications ...

raw ingredients

chemical and microbiological standards

standard of appearance, finish, shape, size, cost

Food companies may use a **quality management system** as their method of quality assurance.

QUALITY MANAGEMENT SYSTEM (QMS)

- improves quality
- builds consumer confidence
- leads to an accreditation certificate
- = British Standards BS5750
- = European Standards 150 9000

Food companies must provide various documents and records:

QMS policy for Fred's Food Company
→ Procedures
→ Specifications
→ Work Instructions
→ Quality records

Staff training

HACCP

Purchases from suppliers

Maintenance of equipment

Customer Complaints

Quality control

Definition
Quality control is part of the quality assurance procedure. It involves checking the quality of food products throughout their manufacture.

QUALITY OF DESIGN

Check against the design specification.

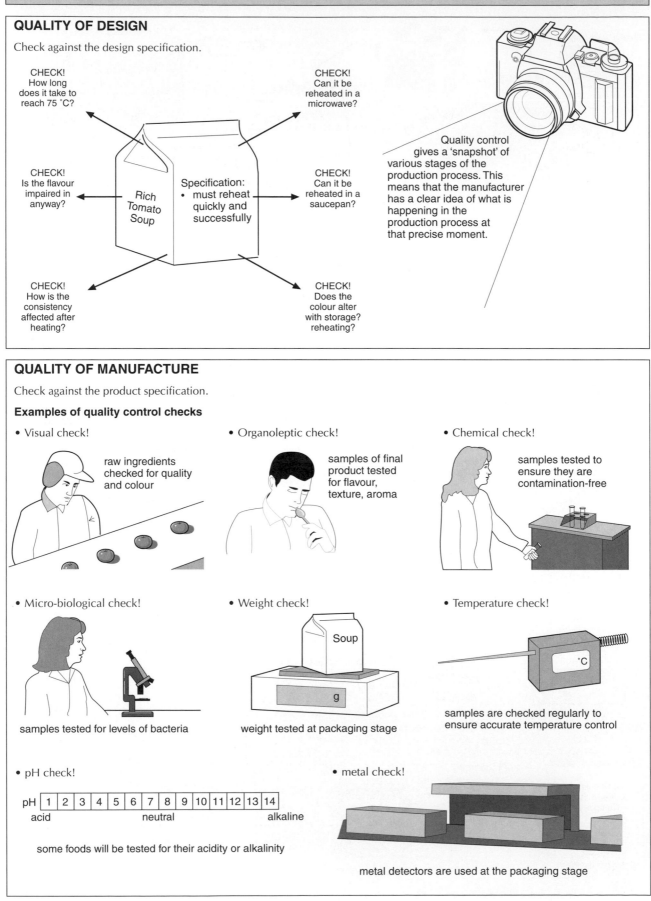

CHECK!
How long does it take to reach 75 °C?

CHECK!
Can it be reheated in a microwave?

CHECK!
Is the flavour impaired in anyway?

Rich Tomato Soup

Specification:
• must reheat quickly and successfully

CHECK!
Can it be reheated in a saucepan?

CHECK!
How is the consistency affected after heating?

CHECK!
Does the colour alter with storage? reheating?

Quality control gives a 'snapshot' of various stages of the production process. This means that the manufacturer has a clear idea of what is happening in the production process at that precise moment.

QUALITY OF MANUFACTURE

Check against the product specification.

Examples of quality control checks

• Visual check!
raw ingredients checked for quality and colour

• Organoleptic check!
samples of final product tested for flavour, texture, aroma

• Chemical check!
samples tested to ensure they are contamination-free

• Micro-biological check!
samples tested for levels of bacteria

• Weight check!
Soup
g
weight tested at packaging stage

• Temperature check!
°C
samples are checked regularly to ensure accurate temperature control

• pH check!

pH	1	2	3	4	5	6	7	8	9	10	11	12	13	14
	acid						neutral							alkaline

some foods will be tested for their acidity or alkalinity

• metal check!

metal detectors are used at the packaging stage

Commercial food production system

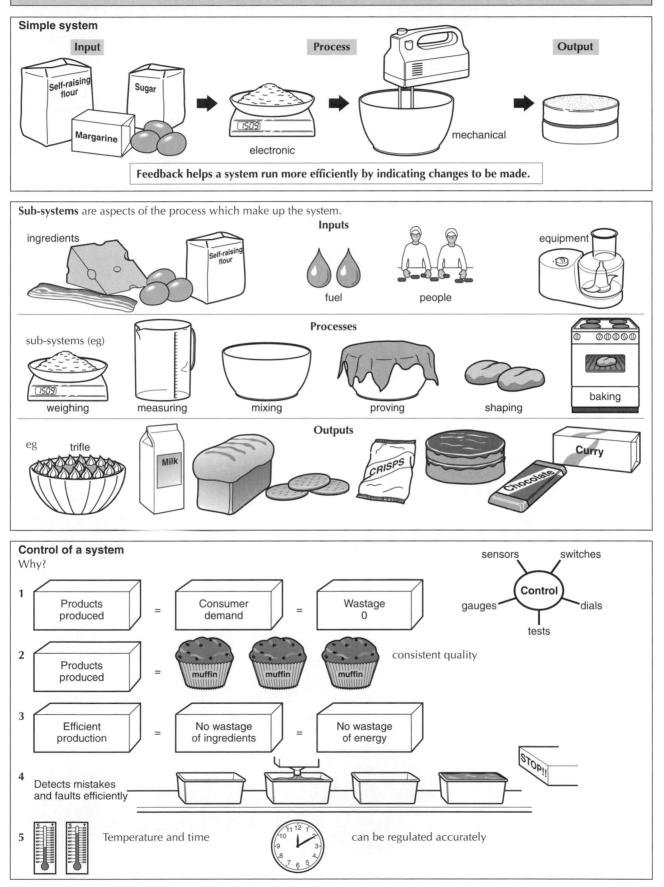

Simple system

| Input | Process | Output |

electronic   mechanical

Self-raising flour · Sugar · Margarine

Feedback helps a system run more efficiently by indicating changes to be made.

Sub-systems are aspects of the process which make up the system.

Inputs

ingredients · Self-raising flour · fuel · people · equipment

Processes

sub-systems (eg) · weighing · measuring · mixing · proving · shaping · baking

Outputs

eg · trifle · Milk · CRISPS · Chocolate · Curry

Control of a system
Why?

sensors · switches · **Control** · gauges · dials · tests

1 Products produced = Consumer demand = Wastage 0

2 Products produced = muffin muffin muffin consistent quality

3 Efficient production = No wastage of ingredients = No wastage of energy

4 Detects mistakes and faults efficiently · STOP!!

5 Temperature and time · can be regulated accurately

CAD/CAM in the food industry

Definition CAD = computer-aided design   CAM = computer-aided manufacture

CAD

Designers can produce 2D or 3D models of new food products ...

... or of packaging using CAD ...

... changes or modifications can be made on screen.

... the designs can be seen by others; in other parts of the office, country or world, if necessary ...

CAM

One person can supervise many operations during food manufacture.

The manufacture of food products can involve the use of computers to control one or more aspect of production. eg computers can control:

oven thermostats

oven temperature and temperature adjustment

moisture content or pH of a product

weighing and measuring

speed of a conveyor belt or other machinery

rate at which an ingredient is added

amount of mixture/coating being used

timing of mixing, kneading, stirring, combining etc.

CIM = computer-integrated manufacture
During manufacture critical adjustments can be made using a computerized control system. Adjustments are registered using **feedback** (see page 29) and made automatically.

Retailers can gather information about food sales using barcodes. Computers interpret the code and provide information which helps with stock rotation, ordering and stock-taking.

‹0040 9193›

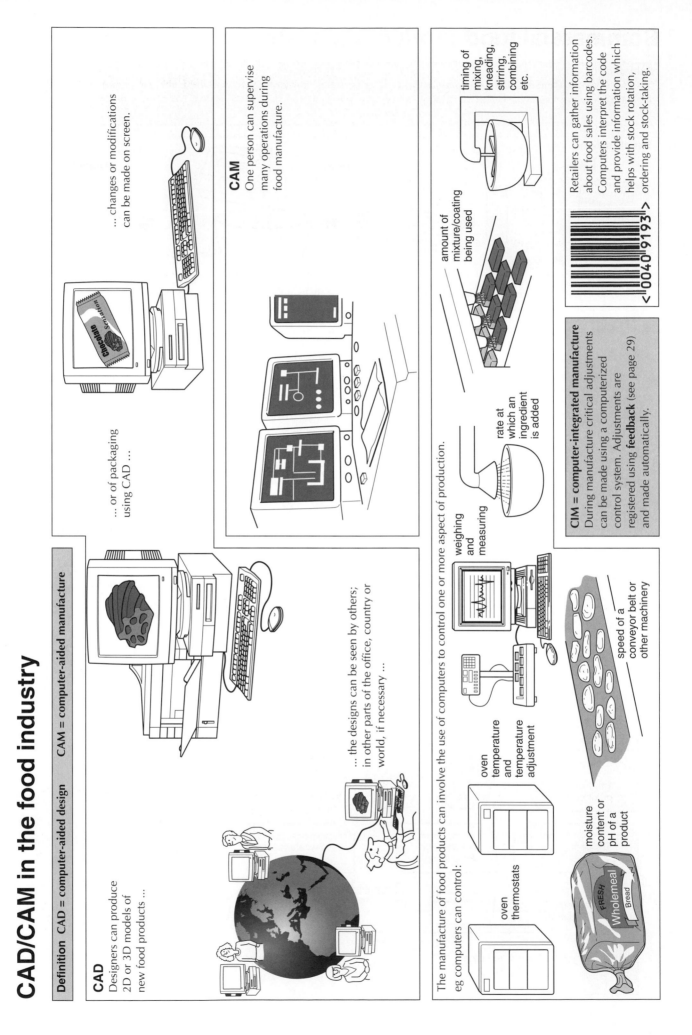

Risk assessment

Definition
Risk assessment is a method of identifying risk factors and assessing the likelihood that a hazard will occur. Food safety is an important risk factor to be considered.

The Food Safety Act 1990 and the Food Safety (General Food Hygiene) Regulations 1995 have led to a new approach in hygiene management.

⇩

• Managing a food production operation must involve the identification of potential hazards.

⇩

• The risks of the hazards occurring are considered.

⇩

• A system for managing them is developed.

Two systems

A hazard = Something which has the potential to cause the consumer harm

HACCP
HAZARD ANALYSIS
CRITICAL CONTROL POINT

developed by the food industry in the early 60's

ASC
ASSURED SAFE CATERING

developed by the Department of Health for caterers

See page 32 for more details.

It is possible to produce a safe food product with HACCP!

1 To establish a HACCP system for a product(s) complete a detailed analysis of the possible hazards:

equipment

Possible hazards may come from the...

food product

production process

2 Once possible hazards have been analyzed, critical control points must be identified:

environment

staff

critical control point (CCP) = a step in a food production system where control is needed to eliminate a hazard or reduce it to a safe level.

3 Appropriate control systems are put in place, eg

4 Control system must be monitored:

2 minutes

75 °C

HACCP

• Temperature control

72 °C ☑

2 mins ☑

food product

cooking process
= 75 °C (+/– 2 °C = tolerance) for 2 minutes

Assured safe catering

Definition
Assured safe catering is a system of managing a catering operation so that potential hazards are identified and the risk controlled.

Step	Hazard		Action
1 Purchase	High risk (ready-to-eat) foods contaminated with food poisoning bacteria/toxins Pate Flan Grated cheese		REPUTABLE SUPPLIER • Specify temperature
2 Receive food	High risk (ready-to-eat) foods contaminated with food poisoning bacteria/toxins Pate Flan Grated cheese		Does it look, smell and feel right? • Check temperature
3 Storage	Growth of food poisoning bacteria toxins		Wrap & label USE BY rotate stock • Check temperatures
4 Preparation	Contamination of high risk (ready-to-eat) foods; growth of food poisoning bacteria		• Wash hands • Use clean equipment • Separate raw and cooked foods
5 Cooking	Survival of food poisoning bacteria		75°C
6 Cooling	Growth of spores/food poisoning bacteria; Production of toxins; Contamination with food poisoning bacteria		• Cool quickly rice • Chill quickly
7 Hot-holding	Growth of food poisoning bacteria; Production of poison by bacteria		• Keep food hot 63°C+
8 Reheating	Survival of food poisoning bacteria		• Reheat to at least 70°C
9 Chilled storage	Growth of food poisoning bacteria		• Check 1–4°C • Check USE BY
10 Serving	Growth of disease-causing bacteria; Production of toxins; Contamination		Cold service – serve immediately to avoid becoming WARM. Hot service – serve immediately to avoid becoming COOL.

Adapted from the Department of Health's Assured Safe Catering CCP poster

Labelling and additives

LABELLING

Consumer

Why are labels useful to me?

1

Baked Beans / Luxury Baked Beans

- help make choices

2

Ingredients:
Beans, water,
tomato purée,
sugar, salt

- help avoid ingredients if necessary

3

Use by = highly perishable food

Best Before = after this date food is not at its best

- help avoid food poisoning

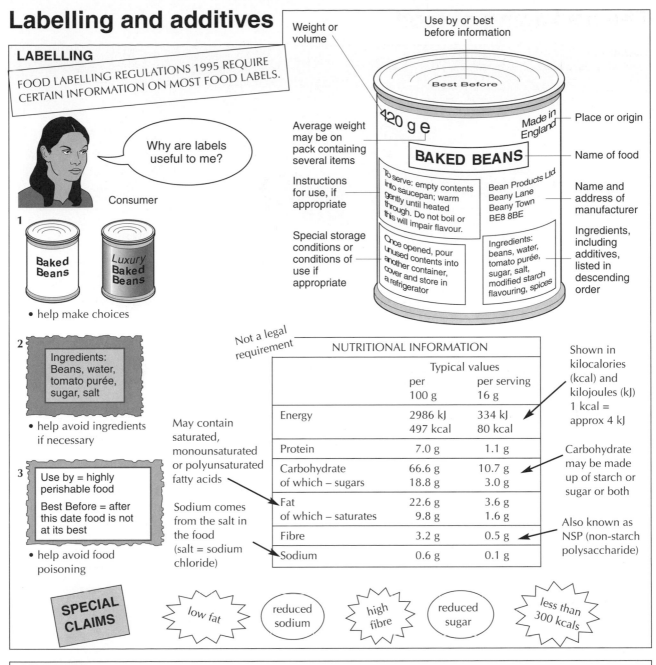

Weight or volume

Use by or best before information

Best Before

420 g e

Made in England — Place or origin

BAKED BEANS — Name of food

Average weight may be on pack containing several items

To serve: empty contents into saucepan; warm gently until heated through. Do not boil or this will impair flavour.

Instructions for use, if appropriate

Special storage conditions or conditions of use if appropriate

Once opened, pour unused contents into another container, cover and store in a refrigerator

Bean Products Ltd
Beany Lane
Beany Town
BE8 8BE — Name and address of manufacturer

Ingredients:
beans, water,
tomato purée,
sugar, salt,
modified starch
flavouring, spices — Ingredients, including additives, listed in descending order

Not a legal requirement

NUTRITIONAL INFORMATION

	Typical values	
	per 100 g	per serving 16 g
Energy	2986 kJ	334 kJ
	497 kcal	80 kcal
Protein	7.0 g	1.1 g
Carbohydrate	66.6 g	10.7 g
of which – sugars	18.8 g	3.0 g
Fat	22.6 g	3.6 g
of which – saturates	9.8 g	1.6 g
Fibre	3.2 g	0.5 g
Sodium	0.6 g	0.1 g

Shown in kilocalories (kcal) and kilojoules (kJ) 1 kcal = approx 4 kJ

May contain saturated, monounsaturated or polyunsaturated fatty acids

Sodium comes from the salt in the food (salt = sodium chloride)

Carbohydrate may be made up of starch or sugar or both

Also known as NSP (non-starch polysaccharide)

SPECIAL CLAIMS

low fat · reduced sodium · high fibre · reduced sugar · less than 300 kcals

ADDITIVES

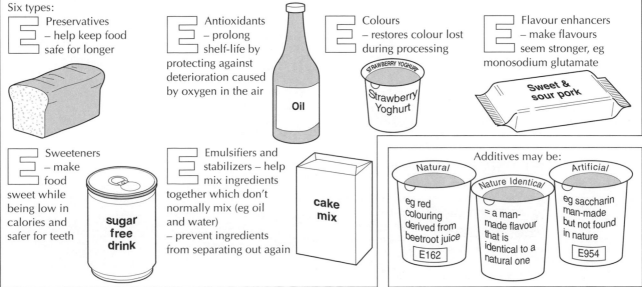

E numbers indicate an additive has been accepted as safe all over the European Union.

Six types:

E Preservatives – help keep food safe for longer

E Antioxidants – prolong shelf-life by protecting against deterioration caused by oxygen in the air

E Colours – restores colour lost during processing

Strawberry Yoghurt

E Flavour enhancers – make flavours seem stronger, eg monosodium glutamate

Sweet & sour pork

E Sweeteners – make food sweet while being low in calories and safer for teeth

sugar free drink

E Emulsifiers and stabilizers – help mix ingredients together which don't normally mix (eg oil and water) – prevent ingredients from separating out again

cake mix

Additives may be:

Natural
eg red colouring derived from beetroot juice
E162

Nature Identical
≈ a man-made flavour that is identical to a natural one

Artificial
eg saccharin man-made but not found in nature
E954

The packaging of food

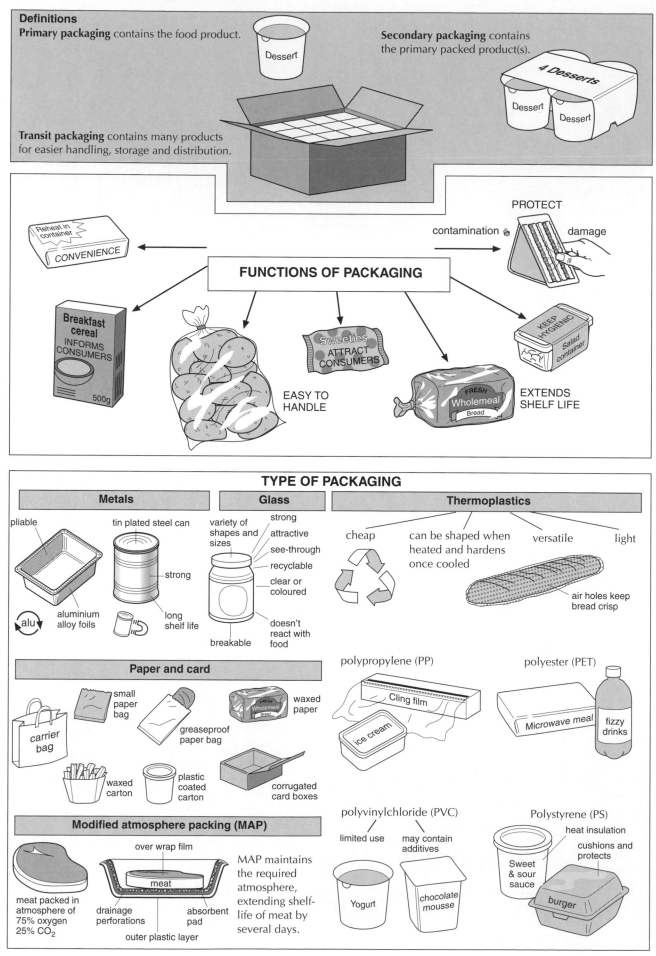

Definitions

Primary packaging contains the food product.

Dessert

Secondary packaging contains the primary packed product(s).

4 Desserts

Dessert   Dessert

Transit packaging contains many products for easier handling, storage and distribution.

FUNCTIONS OF PACKAGING

Reheat in container
CONVENIENCE

Breakfast cereal
INFORMS CONSUMERS
500g

EASY TO HANDLE

Sweeties
ATTRACT CONSUMERS

FRESH
Wholemeal
Bread
EXTENDS SHELF LIFE

PROTECT
contamination   damage

KEEP HYGIENIC
Salad container

TYPE OF PACKAGING

Metals

pliable

tin plated steel can

strong

aluminium alloy foils

alu

long shelf life

Glass

variety of shapes and sizes

strong

attractive

see-through

recyclable

clear or coloured

breakable

doesn't react with food

Thermoplastics

cheap   can be shaped when heated and hardens once cooled   versatile   light

air holes keep bread crisp

Paper and card

carrier bag

small paper bag

greaseproof paper bag

FRESH Wholemeal Bread
waxed paper

waxed carton

plastic coated carton

corrugated card boxes

polypropylene (PP)

Cling film

ice cream

polyester (PET)

Microwave meal   fizzy drinks

Modified atmosphere packing (MAP)

over wrap film

meat

meat packed in atmosphere of 75% oxygen 25% CO_2

drainage perforations

absorbent pad

outer plastic layer

MAP maintains the required atmosphere, extending shelf-life of meat by several days.

polyvinylchloride (PVC)

limited use   may contain additives

Yogurt

chocolate mousse

Polystyrene (PS)

heat insulation

cushions and protects

Sweet & sour sauce

burger

Packaging and the environment

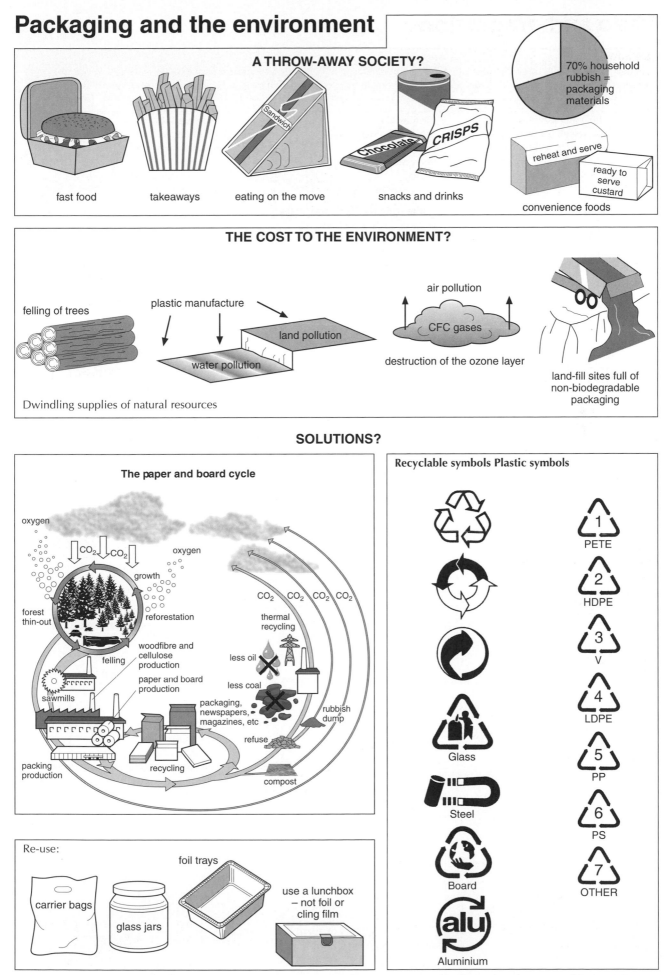

A THROW-AWAY SOCIETY?

fast food takeaways eating on the move snacks and drinks

Sandwich Chocolate CRISPS

70% household rubbish = packaging materials

reheat and serve
ready to serve custard

convenience foods

THE COST TO THE ENVIRONMENT?

felling of trees

plastic manufacture

water pollution

land pollution

air pollution

CFC gases

destruction of the ozone layer

land-fill sites full of non-biodegradable packaging

Dwindling supplies of natural resources

SOLUTIONS?

The paper and board cycle

oxygen

CO_2 CO_2

oxygen

growth

forest thin-out

reforestation

woodfibre and cellulose production

felling

sawmills

paper and board production

packing production

recycling

packaging, newspapers, magazines, etc

refuse

thermal recycling

less oil

less coal

CO_2 CO_2 CO_2 CO_2

rubbish dump

compost

Re-use:

carrier bags

glass jars

foil trays

use a lunchbox – not foil or cling film

Recyclable symbols Plastic symbols

Glass

Steel

Board

alu
Aluminium

1 PETE
2 HDPE
3 V
4 LDPE
5 PP
6 PS
7 OTHER

Taken from the 'The Essential Packaging Solution' PRO Carbon Marketing Bureau, 3 The Plain, Thornbury, Bristol BS12 2AG

Ratio and proportion

Definitions
Ratio = the relationship between the quantities of ingredients in a recipe, eg fat to flour in choux pastry is 2 : 3.
Proportion = the quantity of ingredients in a recipe in relation to others, eg creamed cake has equal proportions of fat to flour.

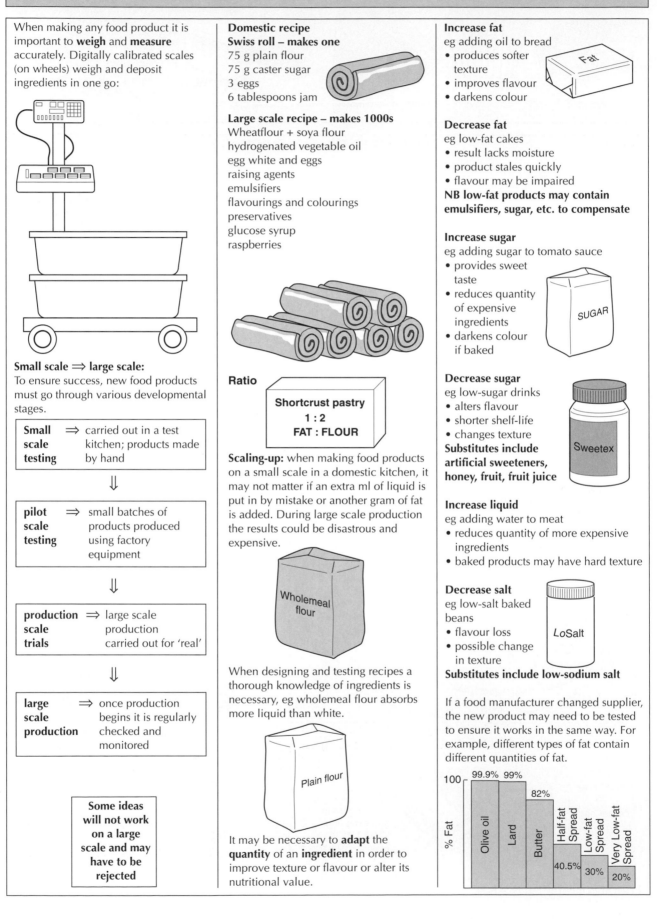

When making any food product it is important to **weigh** and **measure** accurately. Digitally calibrated scales (on wheels) weigh and deposit ingredients in one go:

Small scale ⟹ large scale:
To ensure success, new food products must go through various developmental stages.

Small scale testing	⟹	carried out in a test kitchen; products made by hand

⟱

pilot scale testing	⟹	small batches of products produced using factory equipment

⟱

production scale trials	⟹	large scale production carried out for 'real'

⟱

large scale production	⟹	once production begins it is regularly checked and monitored

Some ideas will not work on a large scale and may have to be rejected

Domestic recipe
Swiss roll – makes one
75 g plain flour
75 g caster sugar
3 eggs
6 tablespoons jam

Large scale recipe – makes 1000s
Wheatflour + soya flour
hydrogenated vegetable oil
egg white and eggs
raising agents
emulsifiers
flavourings and colourings
preservatives
glucose syrup
raspberries

Ratio

Shortcrust pastry
1 : 2
FAT : FLOUR

Scaling-up: when making food products on a small scale in a domestic kitchen, it may not matter if an extra ml of liquid is put in by mistake or another gram of fat is added. During large scale production the results could be disastrous and expensive.

Wholemeal flour

When designing and testing recipes a thorough knowledge of ingredients is necessary, eg wholemeal flour absorbs more liquid than white.

Plain flour

It may be necessary to **adapt** the **quantity** of an **ingredient** in order to improve texture or flavour or alter its nutritional value.

Increase fat
eg adding oil to bread
• produces softer texture
• improves flavour
• darkens colour

Fat

Decrease fat
eg low-fat cakes
• result lacks moisture
• product stales quickly
• flavour may be impaired
NB low-fat products may contain emulsifiers, sugar, etc. to compensate

Increase sugar
eg adding sugar to tomato sauce
• provides sweet taste
• reduces quantity of expensive ingredients
• darkens colour if baked

SUGAR

Decrease sugar
eg low-sugar drinks
• alters flavour
• shorter shelf-life
• changes texture
Substitutes include artificial sweeteners, honey, fruit, fruit juice

Sweetex

Increase liquid
eg adding water to meat
• reduces quantity of more expensive ingredients
• baked products may have hard texture

Decrease salt
eg low-salt baked beans
• flavour loss
• possible change in texture
Substitutes include low-sodium salt

LoSalt

If a food manufacturer changed supplier, the new product may need to be tested to ensure it works in the same way. For example, different types of fat contain different quantities of fat.

% Fat					
100	99.9%	99%			
	Olive oil	Lard	82% Butter	Half-fat Spread 40.5%	Low-fat Spread 30% Very Low-fat Spread 20%

Combining ingredients

During food manufacture ingredients may need to be **combined**. Ingredients interact with each other in different ways. In order to design, develop and manufacture quality food products it is important to know how foods will react.

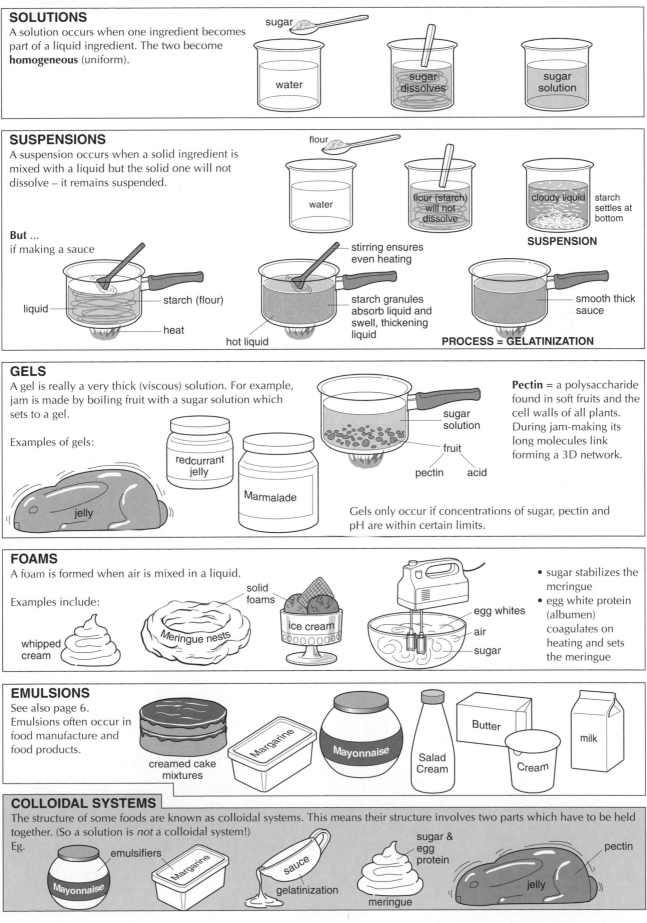

SOLUTIONS

A solution occurs when one ingredient becomes part of a liquid ingredient. The two become **homogeneous** (uniform).

sugar

water

sugar dissolves

sugar solution

SUSPENSIONS

A suspension occurs when a solid ingredient is mixed with a liquid but the solid one will not dissolve – it remains suspended.

flour

water

flour (starch) will not dissolve

cloudy liquid — starch settles at bottom

SUSPENSION

But ...
if making a sauce

liquid — starch (flour)

heat

stirring ensures even heating

hot liquid

starch granules absorb liquid and swell, thickening liquid

smooth thick sauce

PROCESS = GELATINIZATION

GELS

A gel is really a very thick (viscous) solution. For example, jam is made by boiling fruit with a sugar solution which sets to a gel.

Examples of gels:

jelly

redcurrant jelly

Marmalade

sugar solution

fruit

pectin   acid

Pectin = a polysaccharide found in soft fruits and the cell walls of all plants. During jam-making its long molecules link forming a 3D network.

Gels only occur if concentrations of sugar, pectin and pH are within certain limits.

FOAMS

A foam is formed when air is mixed in a liquid.

Examples include:

whipped cream

solid foams

Meringue nests

ice cream

egg whites

air

sugar

- sugar stabilizes the meringue
- egg white protein (albumen) coagulates on heating and sets the meringue

EMULSIONS

See also page 6. Emulsions often occur in food manufacture and food products.

creamed cake mixtures

Margarine

Mayonnaise

Salad Cream

Butter

Cream

milk

COLLOIDAL SYSTEMS

The structure of some foods are known as colloidal systems. This means their structure involves two parts which have to be held together. (So a solution is *not* a colloidal system!)
Eg.

Mayonnaise — emulsifiers

Margarine

sauce — gelatinization

sugar & egg protein

meringue

jelly — pectin

Types and use of industrial equipment

The food industry use a wide range of equipment during food production. Much of it will be familiar in shape and function but it is likely to be a great deal larger than you are used to. Other types of equipment will be unique to the food industry. You will need to become familiar with industrial equipment so you can show how your own food product(s) could be made in industry. You may also need to recognize it in an exam. Remember, different food products can be made using the same equipment (just as a food processor can be used for making cakes, biscuits, soups, pastry, etc.)

Computerized scales ensure accurate weighing.

A centrifuge is like a huge spin drier which can separate liquid from solid parts. It is used during the production of Quorn (see page 40).

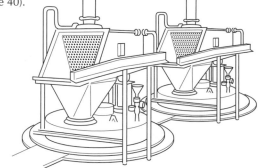

A mandolin is used to slice or cut food products evenly and consistently. It would be used to divide a cake or slice the top to ensure a flat surface before icing.

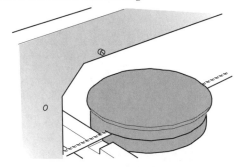

Depositors are used to fill containers with a measured amount of food product. This is likely to be semi-solid, liquid, viscous or aerated (like a mousse).

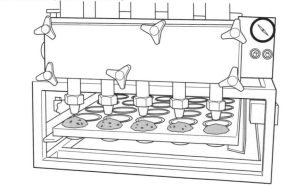

A boiling vat is a large container used for cooking food products such as soup.

Bench or floor standing mixers are large food processors and are likely to be seen in bakeries. The mixing action of the models shown here is known as 'planetary' which ensures consistent mixing.

Date stamping may occur automatically as a product passes along a conveyor belt. First the container is filled then sealed and finally date stamped so the whole process need only take seconds.

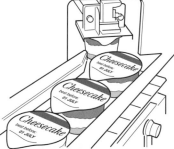

A deck oven range allows batches of products to be cooked simultaneously. This is useful for shops where bread is baked on the premises. The computer controls ensure perfectly baked bread.

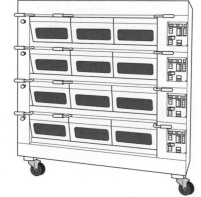

Using micro-organisms to make yoghurt

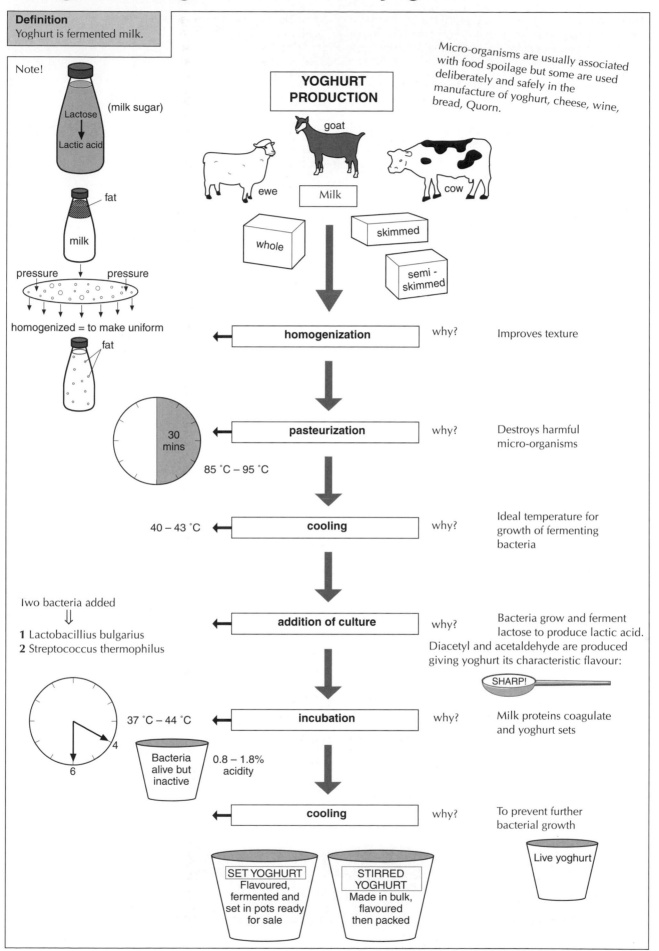

Definition
Yoghurt is fermented milk.

Micro-organisms are usually associated with food spoilage but some are used deliberately and safely in the manufacture of yoghurt, cheese, wine, bread, Quorn.

Note!

Lactose (milk sugar)
↓
Lactic acid

fat

milk

pressure   pressure

homogenized = to make uniform

fat

YOGHURT PRODUCTION

goat

ewe   Milk   cow

whole   skimmed   semi-skimmed

homogenization   why?   Improves texture

30 mins

pasteurization   why?   Destroys harmful micro-organisms

85 °C – 95 °C

40 – 43 °C   **cooling**   why?   Ideal temperature for growth of fermenting bacteria

Two bacteria added
⇓
1 Lactobacillius bulgarius
2 Streptococcus thermophilus

addition of culture   why?   Bacteria grow and ferment lactose to produce lactic acid.

Diacetyl and acetaldehyde are produced giving yoghurt its characteristic flavour:

SHARP!

37 °C – 44 °C   **incubation**   why?   Milk proteins coagulate and yoghurt sets

Bacteria alive but inactive   0.8 – 1.8% acidity

cooling   why?   To prevent further bacterial growth

Live yoghurt

SET YOGHURT
Flavoured, fermented and set in pots ready for sale

STIRRED YOGHURT
Made in bulk, flavoured then packed

Using micro-organisms to make Quorn

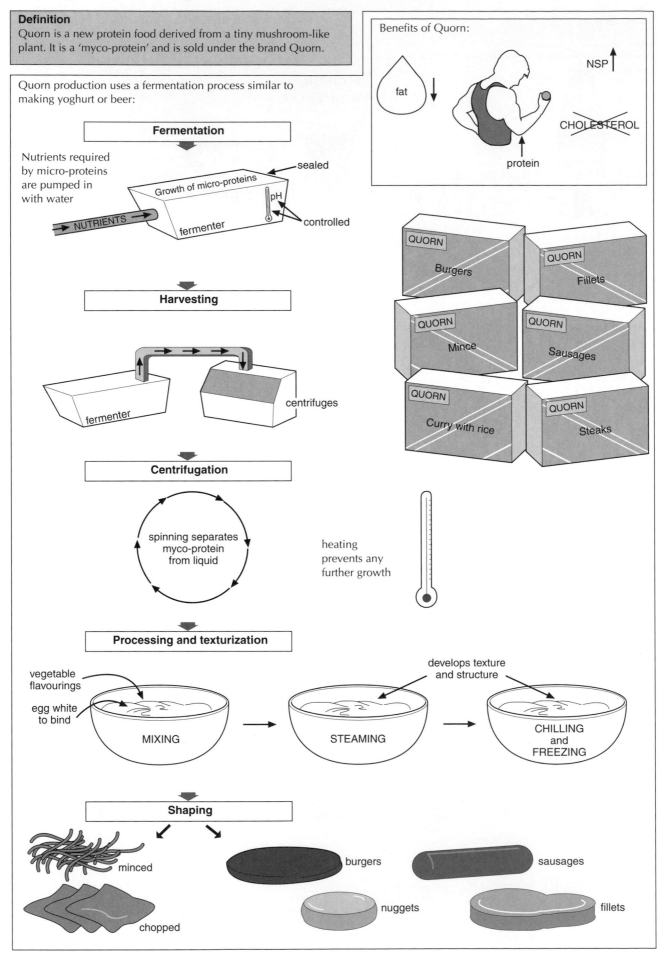

Definition
Quorn is a new protein food derived from a tiny mushroom-like plant. It is a 'myco-protein' and is sold under the brand Quorn.

Benefits of Quorn:

fat ↓

NSP ↑

CHOLESTEROL

protein

Quorn production uses a fermentation process similar to making yoghurt or beer:

Fermentation

Nutrients required by micro-proteins are pumped in with water

sealed

Growth of micro-proteins

NUTRIENTS

pH

fermenter

controlled

QUORN Burgers

QUORN Fillets

QUORN Mince

QUORN Sausages

QUORN Curry with rice

QUORN Steaks

Harvesting

fermenter

centrifuges

Centrifugation

spinning separates myco-protein from liquid

heating prevents any further growth

Processing and texturization

vegetable flavourings

egg white to bind

MIXING

STEAMING

develops texture and structure

CHILLING and FREEZING

Shaping

minced

chopped

burgers

nuggets

sausages

fillets

Presentation of a project

The way you present your coursework is important because it provides the first impression of your work. A well presented project gives a positive impact.

Make some decisions about presentation before you start.

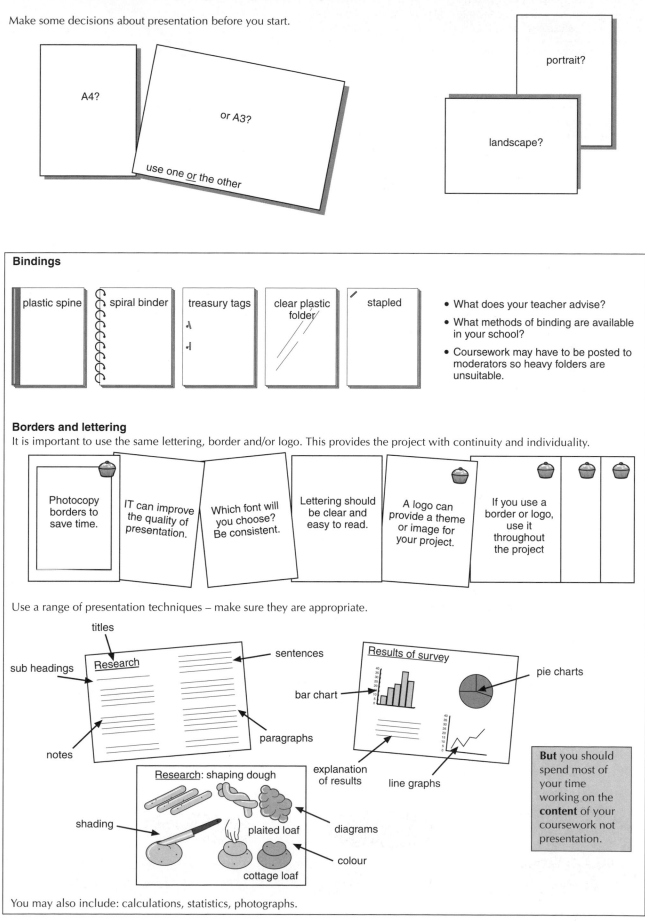

A4?

or A3?

use one or the other

portrait?

landscape?

Bindings

plastic spine | spiral binder | treasury tags | clear plastic folder | stapled

- What does your teacher advise?
- What methods of binding are available in your school?
- Coursework may have to be posted to moderators so heavy folders are unsuitable.

Borders and lettering

It is important to use the same lettering, border and/or logo. This provides the project with continuity and individuality.

Photocopy borders to save time.

IT can improve the quality of presentation.

Which font will you choose? Be consistent.

Lettering should be clear and easy to read.

A logo can provide a theme or image for your project.

If you use a border or logo, use it throughout the project

Use a range of presentation techniques – make sure they are appropriate.

titles

sentences

sub headings

Research

paragraphs

notes

Results of survey

bar chart

pie charts

explanation of results

line graphs

Research: shaping dough

shading

plaited loaf

diagrams

colour

cottage loaf

But you should spend most of your time working on the **content** of your coursework not presentation.

You may also include: calculations, statistics, photographs.

Choosing and organizing a project

You may be given a coursework task, you may choose from a selection or you may have to devise your own. Examples of coursework tasks are shown on page 71.

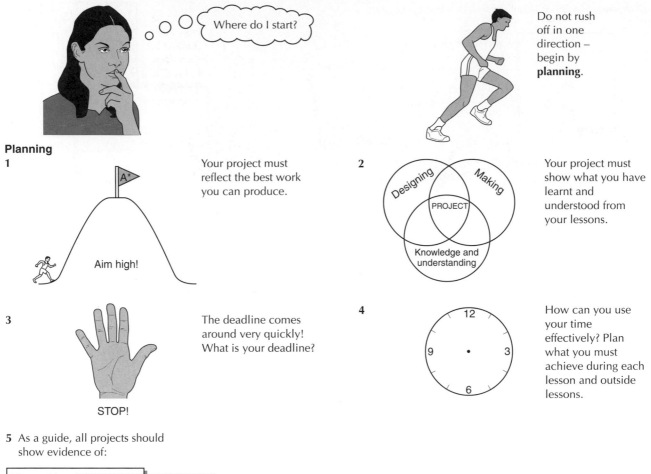

Where do I start?

Do not rush off in one direction – begin by **planning**.

Planning

1 Your project must reflect the best work you can produce.

Aim high!

2 Your project must show what you have learnt and understood from your lessons.

Designing · Making · PROJECT · Knowledge and understanding

3 The deadline comes around very quickly! What is your deadline?

STOP!

4 How can you use your time effectively? Plan what you must achieve during each lesson and outside lessons.

5 As a guide, all projects should show evidence of:

The problem being tackled · Research · Analysis · Specification · Generation of ideas · Development of solution · Planning and production · Evaluation

6 Your project should be presented in a logical order. This needs to be planned in advance:

1 Front cover	2 Problem or need identified	3 RESEARCH: existing products	4 RESEARCH: survey	5 RESEARCH: information gathered
6 Analysis of research: results and conclusions	7 General specification	8 Initial ideas	9 Development of ideas	10 Testing and evaluating an idea
11 Further development	12 Testing and evaluating an idea	13 Product specification	14 Planning production	15 Testing production plan
16 Identifying CCPs	17 Planning a system of HACCP	18 Testing to ensure quality	19 How product can be produced industrially	20 Testing using target audience
21 Further development of idea	22 Final idea Packaging suggestions	23 Testing final idea	24 Sensory evaluation	25 Final evaluation

Showing evidence of industrial practice

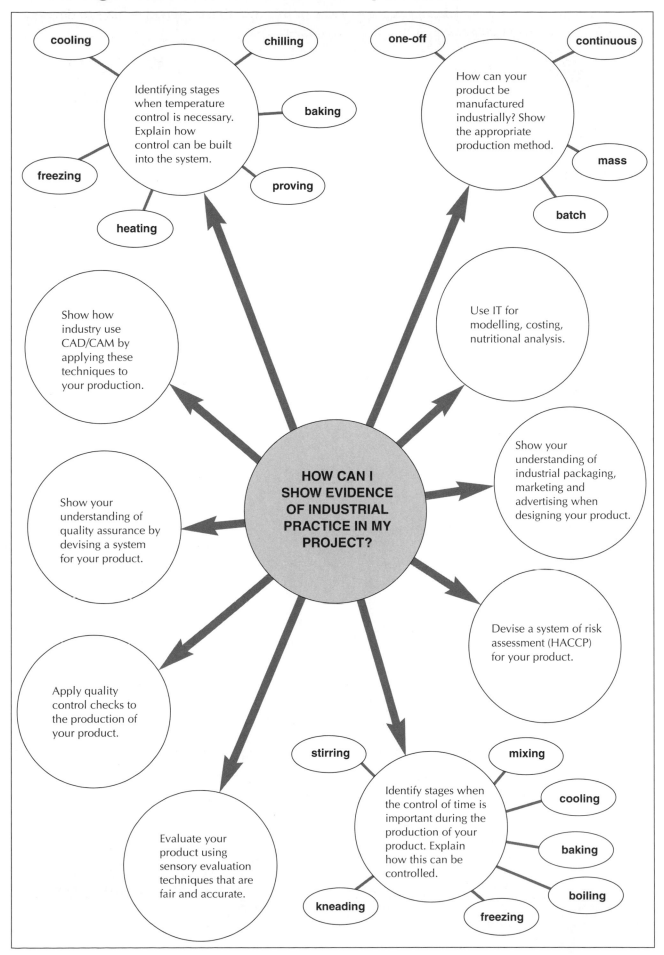

cooling

chilling

one-off

continuous

Identifying stages when temperature control is necessary. Explain how control can be built into the system.

baking

How can your product be manufactured industrially? Show the appropriate production method.

mass

freezing

proving

batch

heating

Show how industry use CAD/CAM by applying these techniques to your production.

Use IT for modelling, costing, nutritional analysis.

HOW CAN I SHOW EVIDENCE OF INDUSTRIAL PRACTICE IN MY PROJECT?

Show your understanding of industrial packaging, marketing and advertising when designing your product.

Show your understanding of quality assurance by devising a system for your product.

Devise a system of risk assessment (HACCP) for your product.

Apply quality control checks to the production of your product.

stirring

mixing

cooling

Evaluate your product using sensory evaluation techniques that are fair and accurate.

Identify stages when the control of time is important during the production of your product. Explain how this can be controlled.

baking

boiling

kneading

freezing

Showing evidence of systems and control

Systems and control is not something to be 'added' to your project. You will show evidence of using systems and control during your project, and perhaps on more than one occasion.

Remember? Look back at pages 29, 31 and 32 to find out about systems and control.

INPUT → PROCESS → OUTPUT

FEEDBACK ← ← ← ←

Opportunities for using systems and control

Production plans   HACCP   Assured Safe catering   Quality control

Production Plan for Fruity Wholemeal Rolls

INPUT

Materials: wholemeal flour, salt, sugar, easy blend yeast, sultanas, apricots

Energy: human; fuel – gas

Equipment: baking tray, bowl, jug, measuring spoons, spoon

PROCESS

1 Weigh ingredients. Keep warm. Grease baking tray. Reheat oven gas mark 6.

HAZARD ANALYSIS and CRITICAL CONTROL Points for Rich Chocolate Dessert

Production stage	Possible hazard	Critical control point
• break chocolate into pieces, place in double saucepan with butter	• chocolate may be out of date	CCP : ensure adequate stock rotation of all dry goods
	• butter may not have been stored at 1–4 °C	CCP : Store all perishable goods in cold store; check thermometer regularly
• beat in egg yolks, mix well	• high risk food – eggs can carry Salmonella bacteria	CCP : ensure quality supplier of eggs; check use by dates; use as fresh as possible

SAFE ASSURED CATERING for SNACK ATTACK Sandwich Bar

Step	Hazard	Action
1 Purchase fresh ingredients	High risk foods may be contaminated	Use good supplier
2 Store goods	Contamination could occur. Food poisoning bacteria could grow	Correct storage of ingredients. Regularly check temperatures

QUALITY CONTROL CHECKS for Marvellous Muffins

Cheese & Mustard Muffins

1 Collect ingredients. Preheat oven gas 6.

2 Weigh ingredients.

Quality Control

Visual check for any defects.

Check scales for accuracy. Weigh carefully.

Starting point

A Food Technology project will start with a problem (task, need, or brief). The aim of the project is to solve that problem successfully.

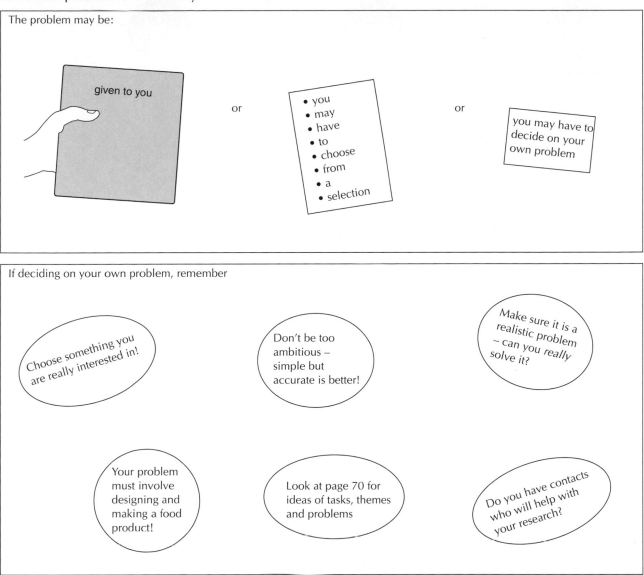

The problem may be:

given to you   or   • you • may • have • to • choose • from • a • selection   or   you may have to decide on your own problem

If deciding on your own problem, remember

Choose something you are really interested in!

Don't be too ambitious – simple but accurate is better!

Make sure it is a realistic problem – can you really solve it?

Your problem must involve designing and making a food product!

Look at page 70 for ideas of tasks, themes and problems

Do you have contacts who will help with your research?

Once you have a problem, read through it several times. In rough, underline the words you feel are particularly important or relevant;

> A food manufacturer requires a low cost range of meals, including a vegetarian option. They are to be sold from the freezer cabinet of local retail outlets.

Ask yourself questions about the problem to be solved:

- What is regarded as 'low cost'?
- What ingredients are economical?
- How many products make a range?
- What size portions are needed?

- What vegetarian ingredients can I use?
- What aspects of packaging, labelling and cost must I consider?
- How will I test my product to ensure it can be frozen?
- What do local retail outlets currently sell?

Asking questions will help you to think through the problem and to write a clear, concise design brief. A design brief is a statement written by you to explain the project:

> To design and make a range of savoury meals which are suitable for freezing

Make sure you record your problem and design brief in your project.

Gathering research information

Before any designing can take place, the problem (task or need) must be thoroughly researched.

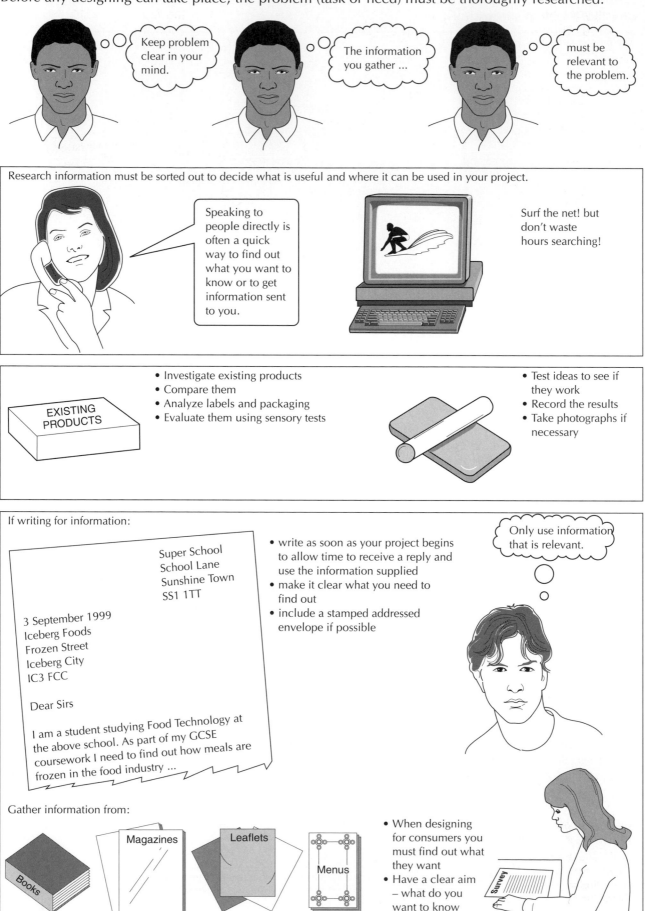

Keep problem clear in your mind.

The information you gather ...

must be relevant to the problem.

Research information must be sorted out to decide what is useful and where it can be used in your project.

Speaking to people directly is often a quick way to find out what you want to know or to get information sent to you.

Surf the net! but don't waste hours searching!

EXISTING PRODUCTS

- Investigate existing products
- Compare them
- Analyze labels and packaging
- Evaluate them using sensory tests

- Test ideas to see if they work
- Record the results
- Take photographs if necessary

If writing for information:

Super School
School Lane
Sunshine Town
SS1 1TT

3 September 1999
Iceberg Foods
Frozen Street
Iceberg City
IC3 FCC

Dear Sirs

I am a student studying Food Technology at the above school. As part of my GCSE coursework I need to find out how meals are frozen in the food industry ...

- write as soon as your project begins to allow time to receive a reply and use the information supplied
- make it clear what you need to find out
- include a stamped addressed envelope if possible

Only use information that is relevant.

Gather information from:

Books   Magazines   Leaflets   Menus

- When designing for consumers you must find out what they want
- Have a clear aim – what do you want to know from consumers?

Survey

Keep your information somewhere safe until you are ready to use it.

Analyzing and using research material logically

During your research you will gather a great deal of information. The information needs to be analyzed. Then it can be used in your project.

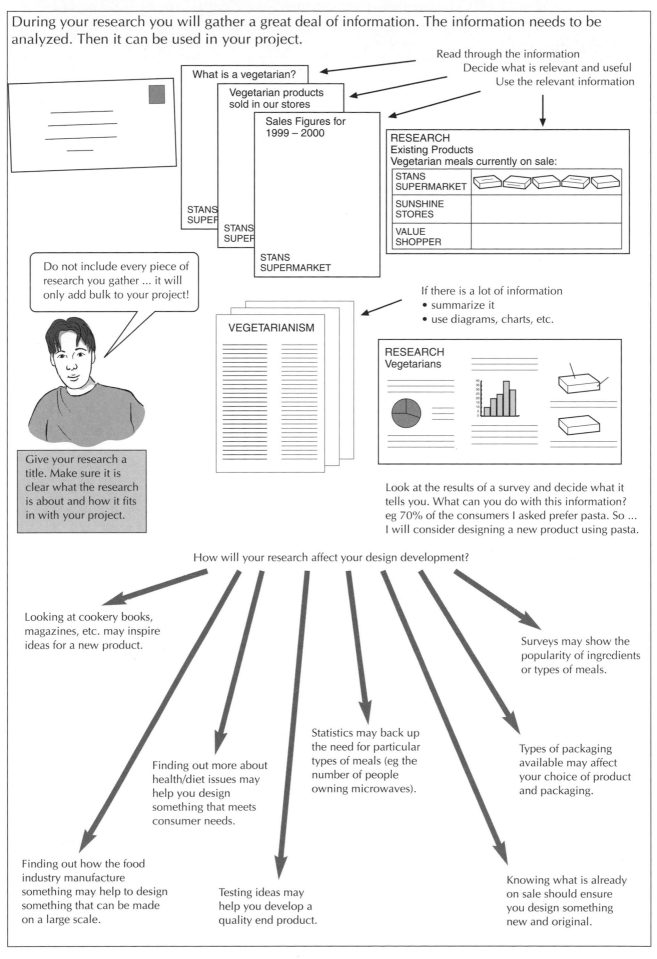

Read through the information
Decide what is relevant and useful
Use the relevant information

What is a vegetarian?

Vegetarian products sold in our stores

Sales Figures for 1999 – 2000

RESEARCH
Existing Products
Vegetarian meals currently on sale:

STANS SUPERMARKET	
SUNSHINE STORES	
VALUE SHOPPER	

Do not include every piece of research you gather ... it will only add bulk to your project!

Give your research a title. Make sure it is clear what the research is about and how it fits in with your project.

VEGETARIANISM

If there is a lot of information
• summarize it
• use diagrams, charts, etc.

RESEARCH
Vegetarians

Look at the results of a survey and decide what it tells you. What can you do with this information? eg 70% of the consumers I asked prefer pasta. So ... I will consider designing a new product using pasta.

How will your research affect your design development?

Looking at cookery books, magazines, etc. may inspire ideas for a new product.

Surveys may show the popularity of ingredients or types of meals.

Finding out more about health/diet issues may help you design something that meets consumer needs.

Statistics may back up the need for particular types of meals (eg the number of people owning microwaves).

Types of packaging available may affect your choice of product and packaging.

Finding out how the food industry manufacture something may help to design something that can be made on a large scale.

Testing ideas may help you develop a quality end product.

Knowing what is already on sale should ensure you design something new and original.

Writing specifications

After analyzing all your research information you can then write a design specification. At a later stage, when you have finalized your product or product range, you will need to write a product specification(s).

DESIGN SPECIFICATION

Your design specification will include information about your product or product range:

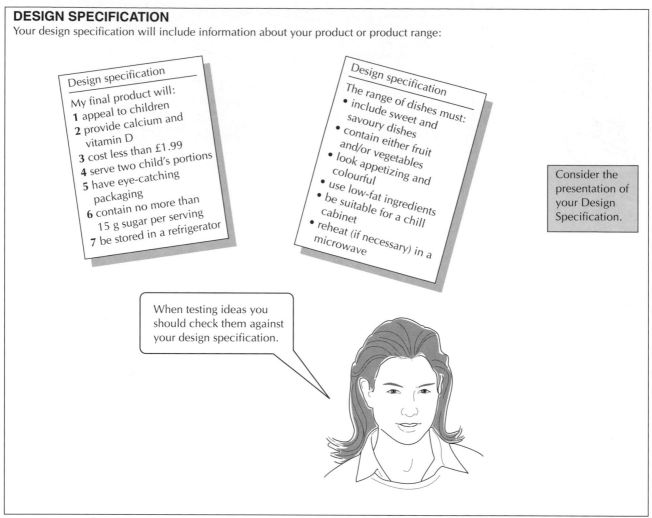

Design specification

My final product will:
1 appeal to children
2 provide calcium and vitamin D
3 cost less than £1.99
4 serve two child's portions
5 have eye-catching packaging
6 contain no more than 15 g sugar per serving
7 be stored in a refrigerator

Design specification

The range of dishes must:
• include sweet and savoury dishes
• contain either fruit and/or vegetables
• look appetizing and colourful
• use low-fat ingredients
• be suitable for a chill cabinet
• reheat (if necessary) in a microwave

Consider the presentation of your Design Specification.

When testing ideas you should check them against your design specification.

PRODUCT SPECIFICATION

A product specification includes details of the final product or one product from a range. Every product has its own product specification. It should provide sufficient information for someone else to make your product exactly as you have designed it. Your final product(s) should be tested against the product specification.

Include:

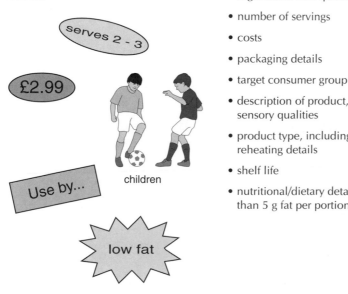

serves 2 - 3

£2.99

Use by...

children

low fat

• ingredients with quantities/proportions

• number of servings

• costs

• packaging details

• target consumer group

• description of product, including sensory qualities

• product type, including storage and reheating details

• shelf life

• nutritional/dietary details (eg less than 5 g fat per portion)

Design ideas

Design ideas can start as rough sketches but gradually they become neater, clearer and more precise as your idea develops. Graphics packages may be used to sketch all or part of your idea.

Always keep your problem, design brief, research and design specification in your mind!

After producing a design specification you should have some ideas for your initial designs.

Initial Ideas for Savoury Bread Snacks

tiny bread sticks

grated cheese inside

bite-size loaves containing herbs

small rolls containing cheese and bacon

Initial ideas : ingredients to flavour bread

nuts

bacon/ham

dried fruits

sun-dried tomatoes

cheese

olives

You don't have to come up with complete ideas in one go – you could consider different aspects such as ingredients, fillings, toppings, finishes, sauces, types of pastry, etc.

PRESENTATION OF DESIGN IDEAS

The dough is rolled out and spread with tomato puree...

Sketches with written descriptions

tomato puree   grated cheese

bread dough

spring onions

rough sketches

annotated sketches

Use more than one method

side

2D

top

shading

grated cheese   tomato puree

bread dough   spring onions

colour

3D

graphics software

Always consider the purpose of design ideas.

Design ideas must **communicate** your ideas to others. They must be presented clearly.

Design ideas: Spiral Bread Snacks

dough containing chopped spring onions

dough spread with tomato puree

rolled up like a swiss roll

grated cheese

Testing design ideas

Design ideas need to be tested to see how well they work.
They may need testing several times until a quality product is produced.

SENSORY EVALUATION

Choose a sensory evaluation method that will find out what you want to know, eg

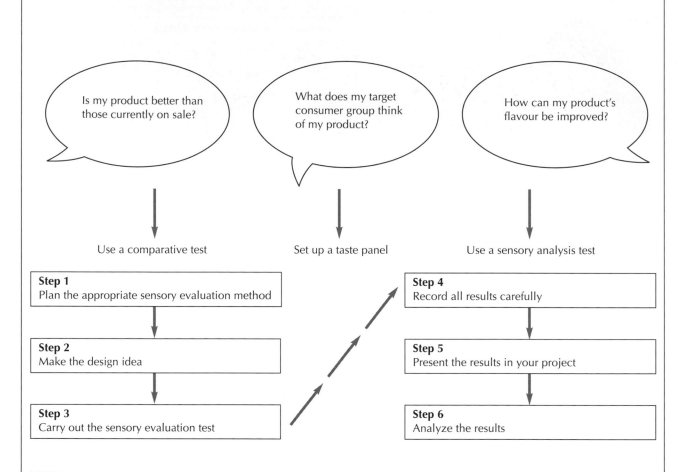

Is my product better than those currently on sale?

What does my target consumer group think of my product?

How can my product's flavour be improved?

Use a comparative test

Set up a taste panel

Use a sensory analysis test

Step 1
Plan the appropriate sensory evaluation method

Step 2
Make the design idea

Step 3
Carry out the sensory evaluation test

Step 4
Record all results carefully

Step 5
Present the results in your project

Step 6
Analyze the results

Testing Design Ideas 1

Sensory Descriptive Profile

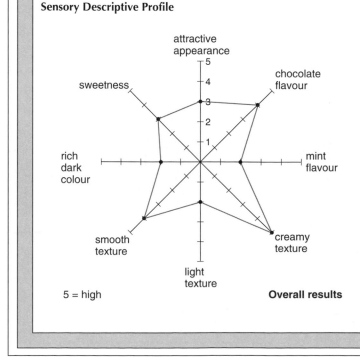

5 = high

Overall results

Chocolate Mint Mousse

The results of my sensory evaluation proved conclusive. People felt the chocolate flavour was strong but the mint flavour was weak. The texture was very creamy and not at all light. I would like to improve on the lightness of the texture. The colour could be improved to make it richer and darker and the appearance could be made more attractive. The level of sweetness was acceptable. In order to improve my product I will need to modify my design idea ...

Evaluating design ideas objectively

When evaluating design ideas you must be **objective**. This means you must base your evaluation on actual results. Your evaluation cannot be biased towards your own opinion or feelings. For example, if you tested a prototype for decorated biscuits and you preferred the ones with the cherries on top but the majority of your testers preferred the chocolate coated biscuits, you mustn't choose the cherry biscuits just because you like them!

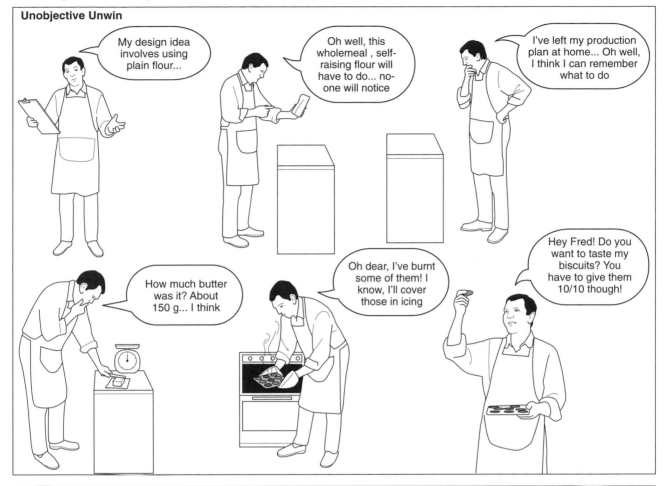

Unobjective Unwin

My design idea involves using plain flour...

Oh well, this wholemeal, self-raising flour will have to do... no-one will notice

I've left my production plan at home... Oh well, I think I can remember what to do

How much butter was it? About 150 g... I think

Oh dear, I've burnt some of them! I know, I'll cover those in icing

Hey Fred! Do you want to taste my biscuits? You have to give them 10/10 though!

Objective Oliver does things differently.

tester 1

- Checks and weighs ingredients accurately
- Follows production plan carefully
- Uses time and temperature control checks
- Notes any problems or difficulties that arise

▲ Has planned how he will evaluate his design ideas
▲ Has asked his testers beforehand
▲ Has prepared evaluation forms for his testers to complete

■ Looks at all the results carefully
■ Totals marks and collates information
■ Prepares an objective evaluation of his design idea

After testing and evaluating a prototype you *might* decide the idea should be rejected. If this is the case, explain *why* it has been rejected.

Modifying design ideas

'Modifying' just means to change or alter something. In this case, it's your design idea. The results from testing and evaluating a design idea may show modification is needed. This means the prototype for the design idea needs to be improved in some way. If testing a prototype for the first time several modifications may be necessary. If a prototype is close to being finalized it may only require 'tweaks' or minor improvements.

WHAT ASPECTS MIGHT REQUIRE MODIFICATION?

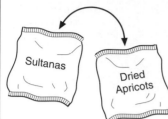

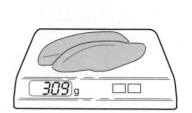

• Change an ingredient

This might be to improve the flavour, texture, aroma or appearance of a product.

• Cutting the cost

Replacing an expensive ingredient for a cheaper one could make the price more acceptable to consumers.

• Increase or decrease portion size

Perhaps consumers felt the portion size was too small or too large?

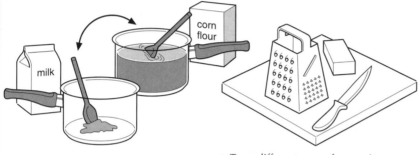

• Alter the nutritional value

You may decide to reduce the fat, sugar or salt, or to increase NSP because consumers felt it wasn't 'healthy' enough.

• Try a different way of preparing an ingredient

Chop instead of grate. Dice instead of slice. Leave unpeeled instead of peeling.

• Reduce or increase temperatures

Accurate temperatures ensure successful products. This applies to all methods of cooking.

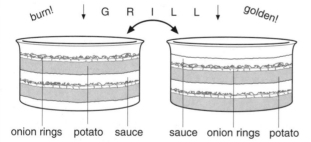

• Put a dish together in a different way

A layered dish might be improved if the layers are altered. Grilling the top, for example, will affect the ingredients on the top layer differently.

• Alter the decoration or garnish

A delicious product may just need to look more eye-catching.

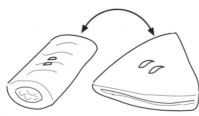

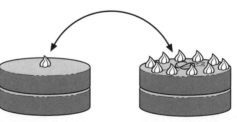

• Try a different shape

You may have designed a tasty product but your testers felt the shape was 'boring'.

• Decrease or increase the time

This may be for baking, grilling, cooling, kneading, mixing, processing, etc.

Use of IT skills

Your project must show evidence of your IT skills. These can be put to good use when carrying out a nutritional analysis of your final product. Or you could use IT to produce some or all of the labelling for your product.

NUTRITIONAL ANALYSIS

It may be essential that you analyze the nutritional content of your product. For example, if you are producing a meal for a diabetic then you must calculate the sugar content. On the other hand, it might just be **desirable** to provide nutritional information on the product's label.

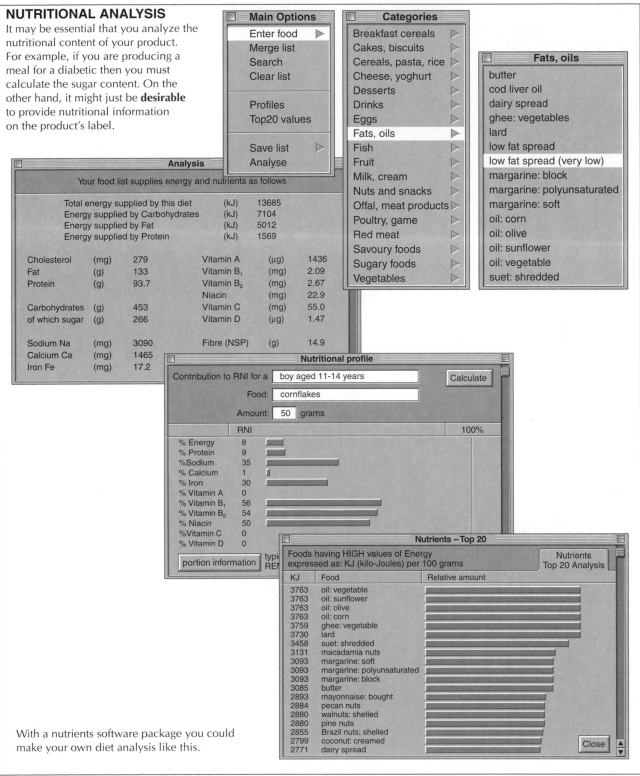

With a nutrients software package you could make your own diet analysis like this.

LABELLING

Make sure you find out whether you **must** show packaging and a label in your project. Check with your teacher.
Otherwise you can produce your own label using a word processing or graphics package.
The following software is useful for producing product labels.
Food in Focus CD Rom (Ridgewell Press Ltd, PO Box 3425, London, SW19 4AX.)

Writing a production plan

A carefully produced production plan is essential if a new product is to be developed successfully. You should be able to give the production plan (along with the product specification) to someone else, and they should make your product exactly as you planned! That is just what happens in the food industry.

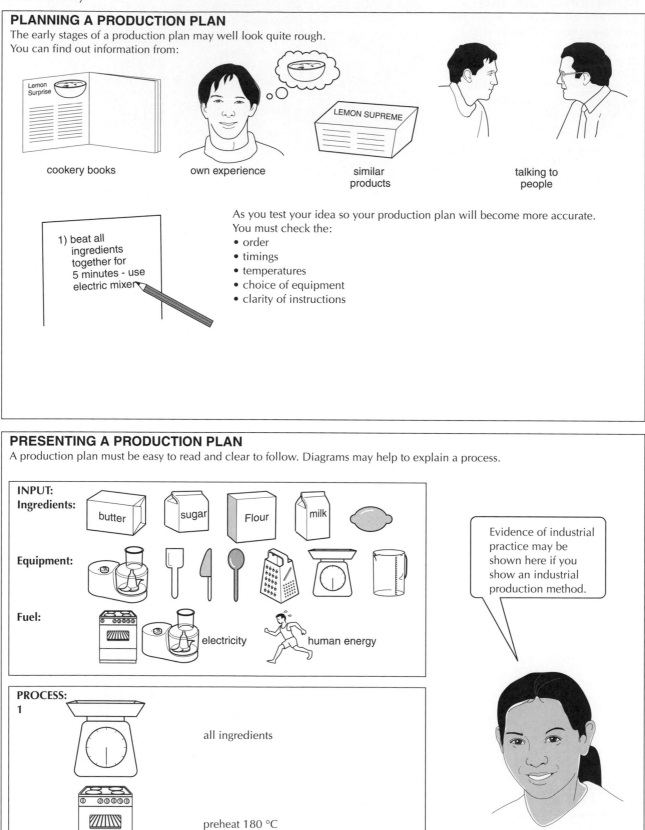

PLANNING A PRODUCTION PLAN

The early stages of a production plan may well look quite rough.
You can find out information from:

cookery books

own experience

similar products

talking to people

1) beat all ingredients together for 5 minutes - use electric mixer

As you test your idea so your production plan will become more accurate.
You must check the:
• order
• timings
• temperatures
• choice of equipment
• clarity of instructions

PRESENTING A PRODUCTION PLAN

A production plan must be easy to read and clear to follow. Diagrams may help to explain a process.

INPUT:
Ingredients: butter   sugar   Flour   milk

Equipment:

Fuel: electricity   human energy

Evidence of industrial practice may be shown here if you show an industrial production method.

PROCESS:
1

all ingredients

preheat 180 °C

Writing production plans showing Quality Control Points

Your aim is to produce a high quality food product.
This can only be achieved if you **plan** for it to be high quality.

PLANNING FOR QUALITY

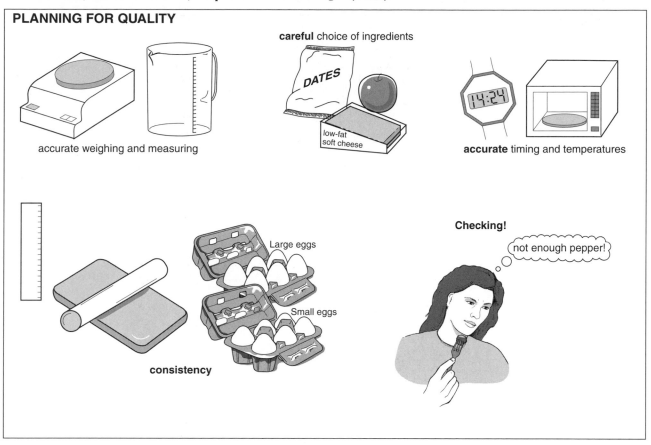

careful choice of ingredients

accurate weighing and measuring

accurate timing and temperatures

Large eggs

Small eggs

consistency

Checking!

not enough pepper!

Showing quality control points when writing production plans

It is important to show how you control and achieve quality during production. This may be done by adding Quality Control Points to your production plan:

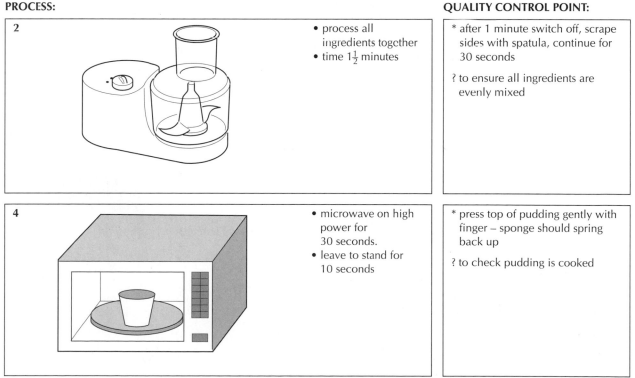

PROCESS:

2
- process all ingredients together
- time 1½ minutes

4
- microwave on high power for 30 seconds.
- leave to stand for 10 seconds

QUALITY CONTROL POINT:

* after 1 minute switch off, scrape sides with spatula, continue for 30 seconds

? to ensure all ingredients are evenly mixed

* press top of pudding gently with finger – sponge should spring back up

? to check pudding is cooked

Writing production plans with evidence of risk assessment

As well as planning for quality it is essential to plan a safe and hygienic production – the end product could not be high quality without it!

ASSURED SAFE CATERING

If your problem involves **catering** for others you may wish to follow the ASC system of risk assessment (see page 32).

HACCP

If your problem involves producing products for **consumers** you may wish to follow the HACCP system of risk assessment (see page 31).

Showing risk assessment when writing production plans

It is possible to include both Quality Control Points and Risk Assessment on one production plan:

Production Plan for Lemony Pud.

INPUT	QCP	HACCP	
		HAZARD	**CCP**

PROCESS

OUTPUT

However, it may be easier and clearer to produce the system of risk assessment separately

Risk Assessment using Hazard Analysis and Critical Control Points

Product: Lemony Pud

Step	Possible hazard	Critical control point
1 Collect ingredients	• incorrect storage • foods out of date	▲ regular check of storage area temperatures

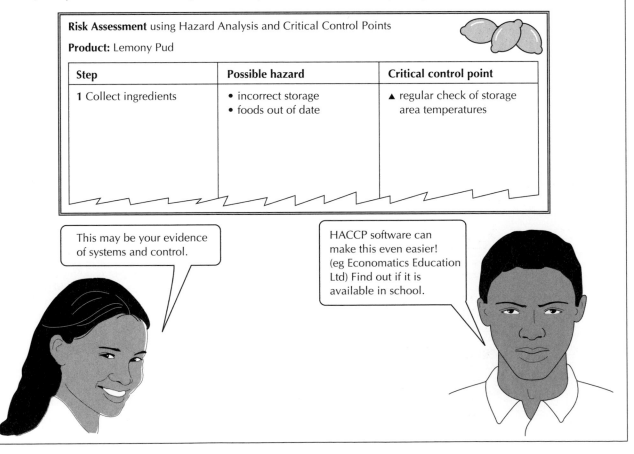

This may be your evidence of systems and control.

HACCP software can make this even easier! (eg Economatics Education Ltd) Find out if it is available in school.

Making successful food products – processes and techniques

Throughout your lessons in Food Technology you will have been building experience and knowledge of food processes and techniques. During your GCSE Food Technology project you must show how competent you have become when working with food as a material.

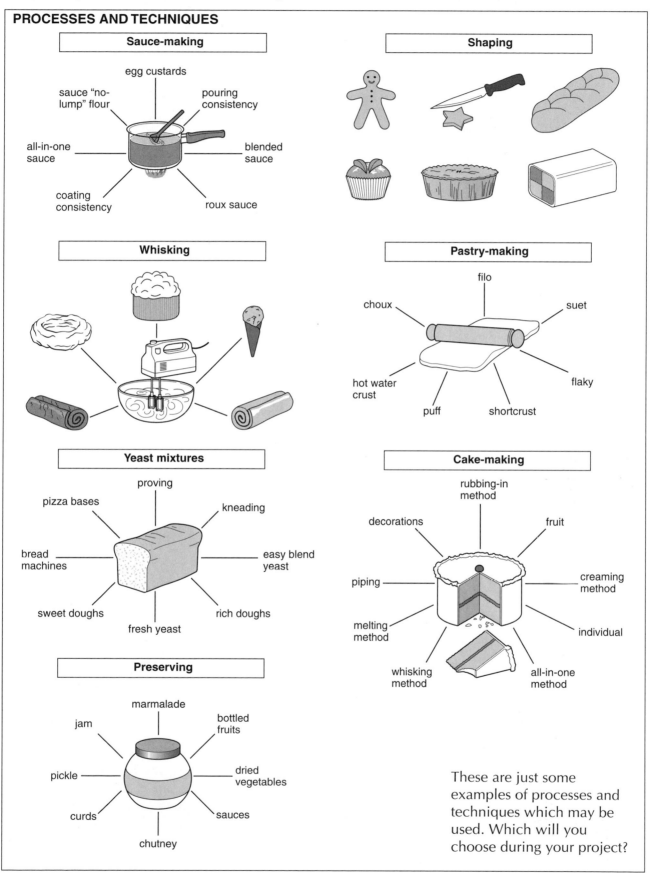

PROCESSES AND TECHNIQUES

Sauce-making
- egg custards
- sauce "no-lump" flour
- pouring consistency
- all-in-one sauce
- blended sauce
- coating consistency
- roux sauce

Shaping

Whisking

Pastry-making
- filo
- choux
- suet
- hot water crust
- flaky
- puff
- shortcrust

Yeast mixtures
- proving
- pizza bases
- kneading
- bread machines
- easy blend yeast
- sweet doughs
- rich doughs
- fresh yeast

Cake-making
- rubbing-in method
- decorations
- fruit
- piping
- creaming method
- melting method
- individual
- whisking method
- all-in-one method

Preserving
- marmalade
- jam
- bottled fruits
- pickle
- dried vegetables
- curds
- sauces
- chutney

These are just some examples of processes and techniques which may be used. Which will you choose during your project?

Making successful food products – equipment

You are unlikely to be able to use industrial equipment when producing your food product. However, you can show which equipment would be used to produce your product on a large scale (see page 38). This would provide some evidence of industrial practice.

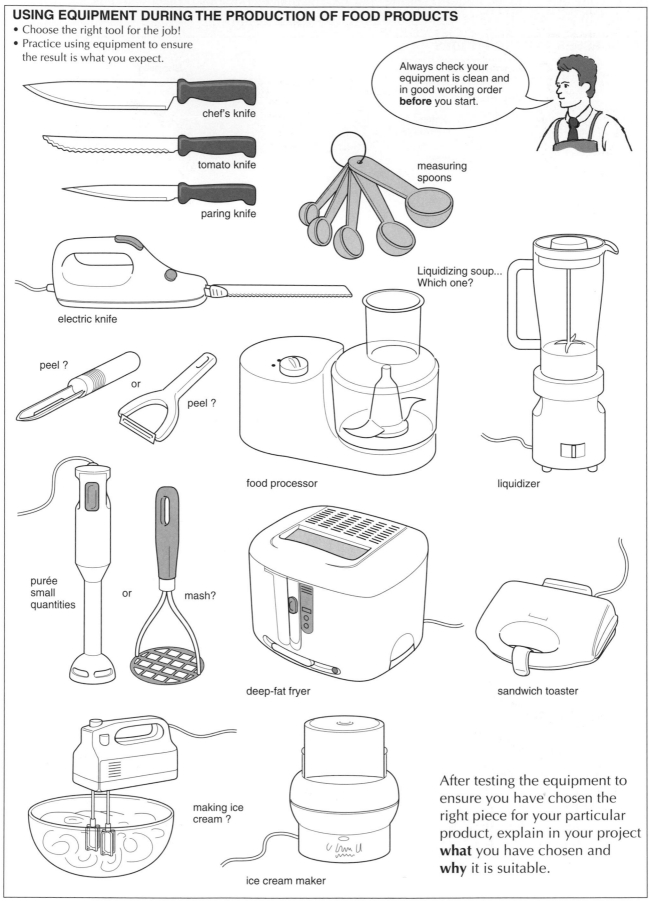

USING EQUIPMENT DURING THE PRODUCTION OF FOOD PRODUCTS
- Choose the right tool for the job!
- Practice using equipment to ensure the result is what you expect.

Always check your equipment is clean and in good working order **before** you start.

chef's knife

tomato knife

paring knife

measuring spoons

electric knife

peel ? or peel ?

Liquidizing soup... Which one?

food processor

liquidizer

purée small quantities or mash?

deep-fat fryer

sandwich toaster

making ice cream ?

ice cream maker

After testing the equipment to ensure you have chosen the right piece for your particular product, explain in your project **what** you have chosen and **why** it is suitable.

Making successful food products – safety and hygiene

As part of your Food Technology project you must identify an appropriate system of risk assessment. However, while you are working you must ensure that **you** always maintain the highest standards of safety and hygiene. Not only will this help to ensure a quality outcome but it will contribute to the marks you are awarded for your coursework.

WORKING SAFELY AND HYGIENICALLY

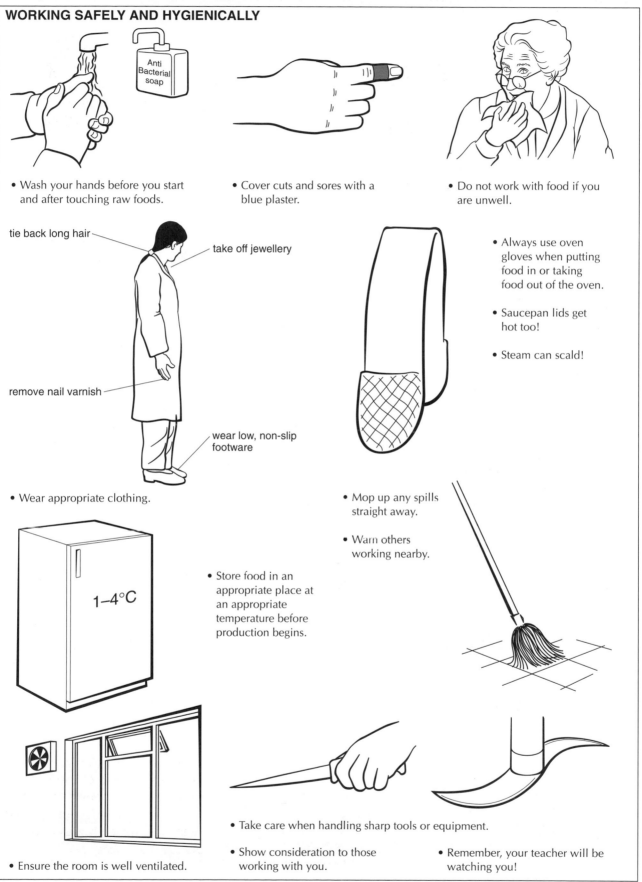

- Wash your hands before you start and after touching raw foods.

- Cover cuts and sores with a blue plaster.

- Do not work with food if you are unwell.

tie back long hair

take off jewellery

remove nail varnish

wear low, non-slip footware

- Wear appropriate clothing.

- Always use oven gloves when putting food in or taking food out of the oven.

- Saucepan lids get hot too!

- Steam can scald!

- Mop up any spills straight away.

- Warn others working nearby.

1–4°C

- Store food in an appropriate place at an appropriate temperature before production begins.

- Ensure the room is well ventilated.

- Take care when handling sharp tools or equipment.

- Show consideration to those working with you.

- Remember, your teacher will be watching you!

Making a quality outcome

The quality of your product must be considered throughout the design and manufacture process.
A good quality product can be achieved through quality designs and quality manufacture.

REMEMBER! Consistency:

Accuracy:

Finally, the quality of your new food product must be the highest you can possibly achieve. Here are some ideas to help you give your final product a quality finish:

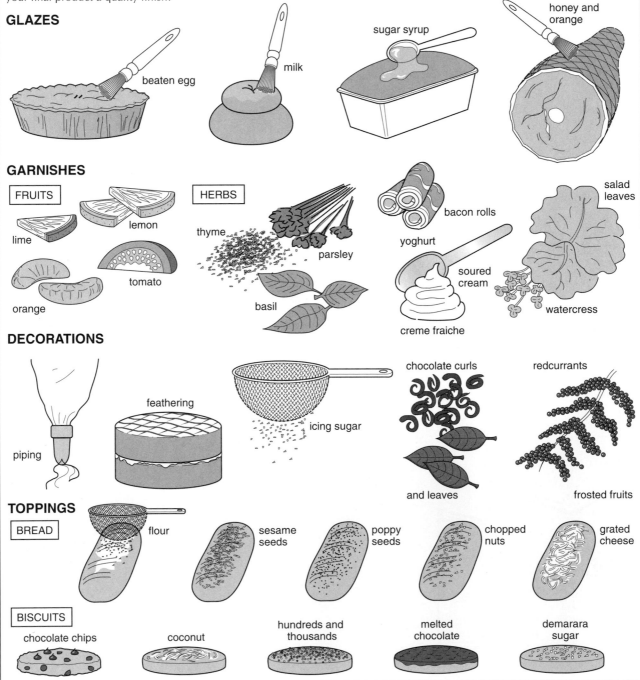

GLAZES

beaten egg

milk

sugar syrup

honey and orange

GARNISHES

FRUITS

lime

lemon

tomato

orange

HERBS

thyme

parsley

basil

bacon rolls

yoghurt

soured cream

creme fraiche

salad leaves

watercress

DECORATIONS

piping

feathering

icing sugar

chocolate curls

and leaves

redcurrants

frosted fruits

TOPPINGS

BREAD

flour

sesame seeds

poppy seeds

chopped nuts

grated cheese

BISCUITS

chocolate chips

coconut

hundreds and thousands

melted chocolate

demarara sugar

Revision questions (pre-examination)

These questions are designed to help you test your knowledge and understanding of Food Technology. They do not represent examination questions. They will make good practice prior to trying examination questions.

1. What is meant by the term 'nutritional value'?
2. Explain the function of protein.
3. Which vitamin assists in the absorption of calcium?
4. Name the four fat-soluble vitamins.
5. Describe two properties of sugar.
6. Explain the process of gelatinization when making a sauce.
7. Proteins are made up of 20 what?
8. Name two sources of high biological value proteins.
9. What happens to protein when it is heated?
10. Name one function of fat.
11. What does 'hydrogenation' mean?
12. Why are emulsifiers important in the production of products such as mayonnaise?
13. Name the natural emulsifiers found in egg yolk.
14. What is meant by a 'primary source' of food?
15. Give one example of a primary source of food.
16. Why are food products made during one-off production more expensive than other production methods?
17. Why are some foods pasteurized?
18. Define 'food spoilage'.
19. Name three types of microbial spoilage.
20. Define 'food poisoning'.
21. Suggest a possible source of the food poisoning bacteria Listeria monocytogenes.
22. What is a safe temperature for a refrigerator?
23. At what temperature is food frozen and stored in industry?
24. What is meant by recipe modification?
25. At what stage does sensory analysis occur during new product development?
26. When might you choose to use a 'difference test'?
27. Name two difference tests.
28. Why is a design specification essential during new product development?
29. Suggest two advantages of using standard components during food manufacture.
30. How is 'feedback' useful during a production system?
31. What does CCP stand for and what is it?
32. When cooking burgers, food poisoning bacteria could remain in the centre causing a potential hazard. What CCP is needed to remove the risk?
33. What is the difference between a 'natural' and a 'nature identical' additive?
34. What is MAP?
35. Suggest three functions of packaging.
36. When flour is mixed with cold water does it become a solution or suspension?
37. What is the process by which starch and liquid become a colloidal system?
38. Pectin is a useful substance during the manufacture of what?
39. Name two examples of solid foams.
40. Micro-organisms are used in the manufacture of which food products? Name five.
41. Quorn is derived from what?
42. What are the nutritional benefits of including Quorn in the diet?

Sample examination questions

FULL/SHORT COURSES – FOUNDATION

Marks awarded are shown in brackets at the end of each question.

1. Look at the pieces of equipment shown below that are found in a kitchen.

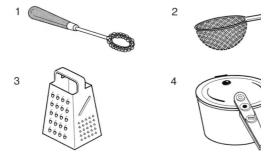

Name each piece of equipment [4]

2. (a) Food handlers who are making sandwiches on a production line must ensure high personal hygiene standards.
   (i) Give **four** hygiene rules which should be observed on the production line. [4]
   (ii) Describe **two** quality control procedures that should be applied to the sandwich production line. [2]

(b) Commercial retailers must ensure that food is supplied in good condition.
   (i) Give **two** ways that butchers avoid contamination in pre-prepared Scotch Eggs. [4]
   (ii) Give **three** examples by which self-service shops maintain the quality of food products on their shelves. [3]

3. Some food products are shown below.

(a) Name **four** different materials that have been used in their packaging. [4]

(b) Choose **one** material from (a) and explain its role in keeping the food fresh. [2]

(c) Choose **one** other material from (a) and explain the role it plays in protecting the food product from damage. [2]

4. Manufacturers sometimes want to develop new fast food products and often employ a market research company to test the market.

(a) Write **five** questions a market research company might ask the consumer when developing a 'dinner for one'. [5]

(b) Having chosen their product, the company must plan a successful product launch.
Give **four** marketing features that they should consider when promoting a new 'dinner for one'. [4]

(c) From the popular convenience foods listed below, choose **two** products and describe their value for a busy consumer.
Cook/chill curry
Prepared pastry case
Fish fingers
Pre-prepared sweets
Frozen apple pie [6]

(d) Describe how convenience foods have contributed to the change in eating habits over the past ten years. [4]

5. The ingredients listed below are for a lasagne produced as a frozen meal.

Lasagne

Meat sauce	Cheese sauce
Minced beef	Plain flour
Onion	Butter
Green pepper	Milk
Mushrooms	Cheddar cheese
Beef stock	
Herbs	Lasagne sheets
Tomatoes	

A manufacturer is to produce a prototype of the lasagne using all the listed ingredients.

(a) Complete the two flow diagrams to explain the main processes used to make lasagne.

1. **Making the Meat Sauce** [5]

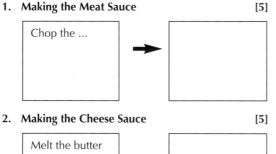

Chop the ...

2. **Making the Cheese Sauce** [5]

Melt the butter

(b) Give the cooking instructions that you would expect to find on the packaging of a frozen lasagne for:
1. **a conventional oven**
2. **a microwave** [6]

6. Bread is often wrapped and packaged for sale.
Part of a label is shown below:

Soft Grain White Bread
White bread with soft grains of Rye and Wheat.
INGREDIENTS
Unbleached, Untreated White Flour, Water, Kibbled Rye, Kibbled Wheat, Yeast, Salt, Vinegar, Soya Flour, Emulsifiers: E471, E472, Vegetable Fat, Flour Improver: Ascorbic Acid (Vitamin C), Folic Acid.

(a) Give **two** reasons why the ingredients on the label are listed in this order. [2]

(b) From the label, name **two** ingredients which are additives. [2]

(c) Give **two** reasons why it is important that the product label lists all the additives used. [2]

(d) This is the nutritional label from a loaf of bread.

NUTRITIONAL INFORMATION	
Average Values per 100 g	
Energy	962 kJ/227 kcal
Protein	7.7 g
Carbohydrate	45.6 g
(of which sugars)	3.4 g
Fat	1.5 g
(of which saturates)	0.4 g
Fibre	3.1 g
Sodium	0.5 g
Vitamin B complex	1.9021 mg
Folic Acid	0.125 mg

(i) Give **two** reasons why the nutritional information given on this label is important. [2]
(ii) Give **one** example for **each** reason. [2]

(e) In addition to the nutritional information, list **five** other points of information which **must** be listed on the packaging of bread products. [5]

7. Bread products can be sold from freezer cabinets.

(a) What is the correct operating temperature range of a freezer cabinet? [2]

(b) How can the correct temperature of the freezer cabinet be checked and maintained? [3]

(c) (i) What causes food to decay? [2]
(ii) Why does putting food in the freezer stop decay? [2]

8. A bakery wishes to extend the range of bread products available to the customer. In the test kitchen a basic recipe for bread is:

Ingredients
375 g bread flour
1 level teaspoon salt
1 sachet (25 g) fast acting yeast
25 g fat
225 ml warm water

By adding or changing ingredients this recipe can be developed to extend the range.

(a) Give **one** idea to develop the basic bread recipe. [1]

(b) Give **two** detailed reasons to explain your idea. [4]

(c) Write out **all** the quantities and the ingredients needed to make your developed recipe. [3]

9. When making the Sweet and Sour Chicken dish, the *Speedy Meals* company carried out a number of control checks.

(a) Explain why all raw ingredients, tins and dried packets are checked before use. [2]

(b) Explain why all ingredients are carefully weighed or cut to a specific size. [2]

(c) Explain why it is necessary to time accurately the preparation and cooking of foods. [2]

10. All food manufacturers carry out Sensory Tests to make sure their products meet the required specification. Explain how a tester would carry out the following tests.

    (a) Triangle Test   [2]

    (b) Ranking Test   [2]

11. Quorn has been developed as an alternative dish to meat and the *Speedy Meals* company is trialling it in its Sweet and Sour dish.

    (a) State **two** reasons why someone might choose the dish made from Quorn instead of meat.   [2]

    (b) State **two** differences you think you would notice when eating the Quorn version of the Sweet and Sour dish compared to the one made with chicken.   [2]

    (c) State **three** nutrients which are found in large amounts in Quorn.   [3]

    (d) Quorn is one of a number of new protein foods. Explain why there is a need to develop these new food sources.   [3]

12. The balance of good health.

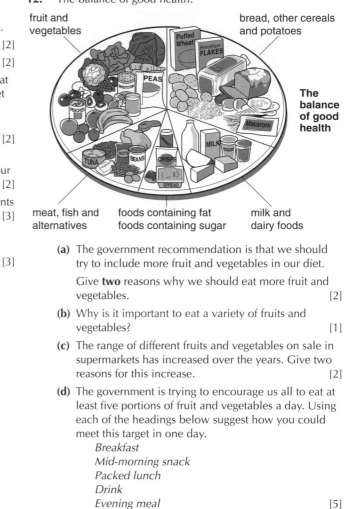

fruit and vegetables — bread, other cereals and potatoes — The balance of good health — meat, fish and alternatives — foods containing fat foods containing sugar — milk and dairy foods

    (a) The government recommendation is that we should try to include more fruit and vegetables in our diet.

         Give **two** reasons why we should eat more fruit and vegetables.   [2]

    (b) Why is it important to eat a variety of fruits and vegetables?   [1]

    (c) The range of different fruits and vegetables on sale in supermarkets has increased over the years. Give two reasons for this increase.   [2]

    (d) The government is trying to encourage us all to eat at least five portions of fruit and vegetables a day. Using each of the headings below suggest how you could meet this target in one day.
          Breakfast
          Mid-morning snack
          Packed lunch
          Drink
          Evening meal   [5]

Sample examination questions

FULL/SHORT COURSES – HIGHER

1. The *Speedy Meals* company has to make sure that its kitchens meets strict hygiene standards. The figure shows a variety of possible problems.

     State **four** possible problems and explain why each of these could be dangerous.   [8]

2. New food processing techniques are continually being developed. Select **two** such techniques and discuss their advantages to **either** the consumer **or** the manufacturer.   [6]

3. A bakery wishes to extend the business and from research has found that household consumption of bread has changed little since 1990.

(The figures represent grams per person per week)

YEAR	1986	1990	1992	1994	1996
White bread	751	623	615	568	540
Brown bread	84	94	94	95	96
Wholemeal bread	18	57	59	60	60
Other/foreign bread	89	99	112	114	120
Total bread	942	873	880	837	816

Using the information in the table above which type of bread has had the greatest increase in sales in the last ten years?   [1]

4. **(a)** The milling process of wheat can be adjusted to the amount and type of flour required from the original grain.

      Type of flour   *Extraction rate*
      Wholemeal   100%
      Wheatmeal   85%
      White   70%

      (i) What does the 'extraction rate' of the flour mean?   [2]

      (ii) White flour has a 70% extraction rate. What is added to this flour during the milling process?   [2]

(iii) Brown/wheatmeal flour is produced using a mixture of wholemeal flour and white flour. Why do manufacturers do this? [3]

(b) Wheat flour is an important ingredient in bread production because of the gluten content.

(i) What is gluten and what are its components? [3]

(ii) Give **four** points to show the importance of gluten in producing a bread product. [8]

(c) Describe an experiment and the expected results that would test the gluten content present in **three** flours. You may use notes and diagrams in your answers. [8]

5. In food production there are usually **three** stages in the system.

What do the following refer to in the production of a bread product?

Give an example in your answer. [6]

INPUT   PROCESS   OUTPUT

6. After a batch production run the following problems have been reported to the test kitchen.

What advice would the food technologist give to correct the following?

(a) After baking a batch of 200 g loaves were found to weigh only 170 g. [2]

(b) Some Chelsea buns are over-baked with a burnt crust. [2]

(c) One batch of bread rolls are small and dense in texture. [2]

(d) The finish quality of unattractive, plain plaited loaves. [2]

7. Commercial companies use a range of methods for preserving food products.

(a) Complete the table below by:
- naming **three** methods of long-term preservation
- giving a food suitable for each chosen example
- giving a reason in each case why the named food is suited to the method of preservation. [9]

Method of preservation	Suitable food product	Reason for choice
eg UHT	Milk	Extends shelf live

(b) Explain why freezing is **not** suitable for preserving tomatoes for salad use. [3]

(c) **(i)** List any **four** of the main processes which occur in the production of UHT (Ultra Heat Treatment) milk. [4]

(ii) Choose **two** of these processes and explain the significance of them. [4]

(iii) Describe **two** effects on the nutritional value of milk products preserved in this way. [2]

8. A food manufacturer wishes to increase the sales of hot main course products in sports centre cafeterias, where the only facility is to reheat food.

(a) **(i)** List **four** criteria that must be considered when designing a suitable food product for this market. [4]

(ii) Choose **two** of your criteria given in **(a)** **(i)** and give detailed reasons for including them in your list. [4]

(b) **(i)** Name a food product which:
- could be served in a cafeteria as a hot main course
- satisfies the dietary needs of sportsmen/women. [1]

(ii) Describe **three** ways by which your answer in **(b)** **(i)** satisfied the dietary needs of sportsmen/women. [6]

(iii) Describe **two** food safety implications of reheating your chosen product. [4]

9. The diagram below shows a simple input/output flow chart for the production of soup.

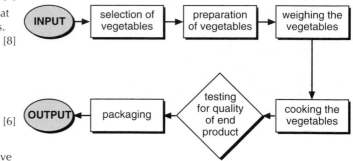

You have been asked by the company to indicate **with reasons** the stages in the production where

(a) human expertise

and

(b) robotic equipment could be used. [10]

10.

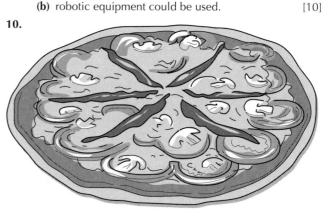

Ingredients	Typical values	Amount per pack approx 4 portions
Wheat flour, Water, Mozzarella medium fat soft cheese, Tomatoes, Olive oil, Salt, Yeast, Oregano.	Energy	760 kcal
	Protein	37.2 g
	Carbohydrates	106.4 g
	(of which sugars)	(12.8 g)
	Fat	20.8 g
	(of which saturates)	(8.0 g)
	Non-starch polysaccharide	9 g
	Sodium	1.0 mg
	Calcium	496.0 mg
	Iron	5.2 mg

(a) How much non-starch polysaccharide is found in one portion of this pizza? [1]

(b) If 18 g of non-starch polysaccharide is the dietary reference value per day what percentage does the amount in one portion contribute to this? [1]

(c) Suggest **two** ways in which you could modify this pizza to ensure that the percentage of non-starch polysaccharide is increased. [2]

(d) How do the performance characteristics of cheese change when baked on a pizza? [6]

11. Evaluate the changes that the Food Safety Act 1990 has had on fast food outlets. [10]

Revision answers

1. The number and type of nutrients provided by a food.
2. Growth, repair and maintenance of body cells and tissues.
3. Vitamin D.
4. Vitamins A, D, E, K.
5. (I) sweetener
   (ii) improves colour
   (iii) preservative
   (iv) retains moisture
   (v) helps incorporate air
   (vi) prevents development of gluten
   (ii) strengthens protein
   (viii) prevents enzymic browning
   (ix) adds crunch/texture
   (x) provides source of food for yeast
   (xi) with pectin, allows jams to set
6. Starch granules and liquid are heated; as starch heats up granules absorb liquid, swell and thicken sauce.
7. Amino acids.
8. Meat, fish, soya beans, cheese, eggs, milk.
9. Protein denatures and coagulates (sets).
10. (i) insulates internal organs
    (ii) concentrated form of energy
    (iii) source of vitamins A, D, E, K
    (iv) essential to structure and function of body cells.
11. The process of altering oils into solids through the addition of hydrogen (eg margarine).
12. Help keep ingredients mixed together (ingredients that would otherwise separate out, eg oil and water).
13. Lecithin.
14. Raw material which become ingredients or secondary sources of food.
15. Wheat grain, sugar beet, sugar cane, maize, coffee beans, tea, rice.
16. Due to personal service, highly skilled craft people, high quality, unique end product.
17. Destroys pathogenic micro-organisms; short-term preservation.
18. Food that appears unpalatable but may not actually be harmful; natural process.
19. Bacteria, moulds, yeast.
20. Occurs when contaminated food has been consumed; food may be contaminated by moulds, viruses, bacteria or their toxins, or chemicals, or naturally toxic plants/animals.

21. Brie, pate, pre-packed salad.
22. 1–4 °C.
23. 29 °C
24. To adapt or change ingredient(s) in type or quality or alter process(es) of the production.
25. Throughout the process of development.
26. To find out if there is a detectable difference between two or more food samples.
27. Triangle test; paired comparison test.
28. It provides details of the product to be designed.
29. (i) ensures consistency (of size, quality, flavour, depth, ingredients)
    (ii) saves time.
30. It helps a system run more efficiently by indicating changes to be made.
31. Critical control point; a step in a food production system where control is needed to eliminate a hazard or reduce it to a safe level.
32. Use food probe to check centre of burger reaches at least 70 °C.
33. Natural = derived from food product; nature identical = man-made but identical to flavour, colour, etc found naturally.
34. Modified atmosphere packaging; maintains required atmosphere to extend a food's shelf life.
35. (i) attract consumer
    (ii) inform consumer
    (iii) hygiene
    (iv) convenience
    (v) extend shelf life
    (vi) protect
    (vii) easy to handle/transport .
36. Suspension.
37. Gelatinization (see question 6).
38. Jam.
39. Ice cream; meringue.
40. (i) yoghurt
    (ii) cheese
    (iii) wine
    (iv) bread
    (v) Quorn
41. A tiny mushroom-like plant.
42. Low in fat, reasonable amounts of NSP, no cholesterol, high quality protein.

Sample examination answers

FULL/SHORT COURSES – FOUNDATION

1. (1) Handheld whisk/whisk/coil whisk
   (2) Sieve
   (3) Grater/multipurpose grater and slicer
   (4) Pressure cooker
2. Detailed explanation to include:
   (a) (i) Ensure hair is covered/tied back; do not cough, etc. over food; cover cuts/grazes with blue plaster; wash hands; no handling rubbish/touching your face; no jewellery; protective clothing (hats, gloves, etc.)
   (ii) Check standard quantities being applied; ensure no cross contamination by using separate cutlery and equipment; ensure storage of ingredients at correct temperature; finished product fully sealed.
   (b) (i) Keep raw meat and processed foods separate; never cut raw meat and processed foods on the same chopping board; use separate cutlery/equipment; prevent food handlers from dealing with raw and processed foodstuffs; store at correct temperature (1 °C and 4 °C); wash hands/wear gloves.
   (ii) Ensure dry foodstuffs are at least 45 cm/18 inches off floor; removing foodstuffs that are past their use by/best before date; remove any 'blown' tins

from stock; rotate stock; storage of food in fridge/freezer for named iterms; temperatures correct for fridge 1–4 °C and freezers –18 °C.

3. (a) Cardboard/waxed cartons or tubs; tin (or can); paper (plain or waxed); foil; plastic clingfilm/polystyrene.

   (b) 1 mark for commenting that packaging prevents contamination of food; 1 mark for specific knowledge about material that relates to keeping food in prime condition (eg sugar kept dry).

   (c) Physical damage or food spoilage ('contamination' only gets a mark once).

4. (a) Questions should relate to: male/female; single/with partner; patterns of buying; price range; main meals of the day/supper dish; importance of nutritional content; healthy eating issues; kJ/kcal content; vegetarian; foreign dish.

   (b) Advertising campaign (TV, magazines, news); leaflet drop to households; money-off tokens; three for the price of one; in-store tasting; prominent place in store; celebrity launch.

   (c) Values include: reduced preparation time; variety of flavours/types; reduced level of skill (pastry case); guaranteed success; rapid meal which can be microwaved or reheated in conventional oven; can be cost effective; little wastage.

   (d) Reasons to include: less time spent in food preparation; more women working therefore less time to prepare food; advances in food technology (give examples); increased freezer ownership; influence of advertising on people's food habits; quick; easy; saves time and fuel; can be kept for emergencies; usually little wastage; often have extra nutrients added.

5. (a) Flow diagrams to include:

   1. Meat Sauce
   Chop onion and pepper
   Slice mushrooms
   Dry fry mince until liquids are clear
   Add onions and pepper and cook for 3–4 minutes
   Add mushrooms and cook for further 2 minutes
   Add tomatoes and beef stock, simmer for 10 minutes

   2. Cheese Sauce
   Melt fat
   Add flour and cook for 2 minutes
   Off heat, add milk, a little at a time
   Return to the heat and bring to the boil
   Allow to thicken and simmer for 2–3 minutes
   Off the heat, add half the grated cheese

   (b) Cooking instructions:

   1. A conventional oven: remove all paper/plastic-based packaging, place on baking tray and cook for 20–25 minutes, heat oven to 200 °C/gas mark 6, check product is cooked thoroughly throughout.

   2. A microwave oven: remove outer packaging, pierce cellophane in several places, heat and stand, timings depend on type/power level of microwave.

6. (a) By law; ingredients must be in descending order of quantity.

   (b) Additives include:
   E472, E472 – help fat and water mix together/holds tiny droplets of oil suspended in water

   Ascorbic acid (vitamin C) – helps in bread production, reduces time necessary for rising/produces greater elasticity in dough

Folic acid – protection against neural tube defect at time of conception and early stages of pregnancy.

   (c) Additives can cause allergies in some people; hyperactivity in children; by law

   (d) (i) Meets legal requirements/provides information for the consumer
       (ii) Legal requirement – date mark/use by date/best before date; name of product; weight/volume; country/place of origin; name and address of manufacturer/EC seller; storage instructions; preparation/cooking instructions; special claims.

       Consumer information – serving suggestions; nutritional information; portion size; cost.

   (e) See answer for **(d) (ii)** – any legal requirement not already mentioned.

7. (a) –18 °C – 29 °C.

   (b) Checked – use of thermometer placed permanently in freezer and regularly checked; use of food probe inserted in cabinet for two minutes; computerized logging gives continuous temperature record; digital display linked to alarm system.

       Maintained – keep doors/lids shut; regular defrosting; thermostat to control temperature range; ice no thicker than 1 cm.

   (c) (i) Food decays naturally and the process is speeded up by enzymes. Harmful decay of food is caused by micro-organisms (bacteria and yeast).
       (ii) Micro-organisms can still grow at 4 °C (fridge temperature) but at temperature of freezer (–18 °C) growth is retarded. Also, water in food is unavailable to micro-organisms. Once thawed, activity resumes and micro-organisms continue to multiply.

8. (a) Ideas to include: change flour to wholemeal/granary; water to milk; additional ingredients to flavour such as dried tomatoes/olives/herbs; suggested a finish (poppy seeds/sesame seeds).

   (b) Explanation may include: dietary factors (increase NSP); target consumer group (children); cultural influence (Italian ingredients).

   (c) Specify ingredients and quantities clearly (eg **strong** flour); recipe must show adaptation and must be a workable recipe (do not miss anything out!)

9. (a) To ensure they are: not damaged in any way/still in date/have not gone off/mouldy.

   (b) Reasons to include:
   Working with a budget/set weight of ingredients; carefully costed to ensure a profit is made; to ensure the correct amount of ingredients are used in each dish; to produce a consistent product; cut to a specific size to ensure even cooking of all ingredients.

   (c) Reasons to include:
   A lot of preparation is computer-controlled to take a certain amount of time so 'x' amount of products can be made in a specified time to ensure thorough cooking of chicken and sauce; to ensure continuity and quality of product (same taste/appearance); to ensure nothing is under- or over-cooked.

10. (a) **Triangle Test** = identify odd one out from total of three samples (full explanation required).

    (b) **Ranking Test** = different samples provided. Tasters put samples in order of preference in relation to colour, taste, texture, odour (full explanation required).

    Codes and symbols used to mark samples.

11. **(a)** Suitable for vegetarians; free from artificial additives; no cholesterol; one third less fat than meat; as much protein as meat; more NSP than meat (**not** just 'healthier').

(b) May lack flavour; may be chewier; may lack colour; different texture to meat.

(c) Protein; carbohydrate; iron; NSP; zinc; biotin.

(d) Other sources of protein are expensive to produce; people want new products to try; manufacturers trying to reduce costs of food; production to maintain a profit; increase in number of vegetarians wanting new products/provides more choice.

12. **(a)** Two explanations from: provide vitamins; minerals/NSP; low in fat/natural source of sugar; source of vegetable protein (**not** just 'healthy').

(b) A variety of fruit and vegetables will provide a balance of nutrients and improve intake of NSP.

(c) Two explanations from: increased demand; wider travel; better advertising; improved transportation; improved storage; multicultural society; increased number of vegetarians; more adventurous recipes tried by consumers; more imported foods.

(d) Five different ideas required, for example:
Breakfast – fruit juice, fruit with cereal/muesli, addition of dried fruit, fruit yoghurt, grapefruit

Snack – fruit, dried fruit in cake, little packets of mixed fruit, fruit drinks

Lunch – salad foods, piece of fruit, salad ingredients in sandwiches

Drink – fruit used to make fresh fruit drinks

Evening meal – any main course or pudding which makes use of fruit and vegetables.

Sample examination answers

FULL/SHORT COURSES – HIGHER

1.

Possible problems	Explanation
Raw/cooked food together	Cross-contamination. Bacterial growth
Food left uncovered	Attract dust, flies. Incorrect storage temperature could lead to bacterial breeding
Wiping hands over nose/throat (sneezing)/cold/cough	Bacterial infection
Cigarette	Ash in food. Stub falls in food. Hand to mouth bacteria
Long hair	Hair may fall in food
Open bin	Attracts vermin/flies/carriers of food poisoning bacteria
Cat	Dirt from paws contains bacteria. Fur in food
Refrigerator overloaded	Temperature not at correct level/bacteria start to breed. Cross contamination from other foods
Mice/rats/vermin/rodents/infestations	Carry bacteria. Eat/gnaw food containers. Fur in food. Droppings in food

2. Technique: accelerated freeze drying
Colour, texture, most of flavour retained; fewer vitamins lost; products last longer than other dried products; used with coffee; complete meals; fish

Technique: cook-chill
Fully prepared, packed and chilled rapidly; only needs reheating by consumer; has led to increase in ready-meal markets both in volume and variety; increases in shelf life of food; saves consumer time.

Technique: irradiation
Reduces number of bacteria; kills insects and pests and increases shelf life of food by delaying enzyme activity. Foods involved are potatoes, onions (reduces sprouting), fruit (slows down ripening therefore can be transported greater distances); kills pests in wheat, rice, spices, meat to kill bacteria and all viruses therefore prolonging shelf life. Only licensed in UK for herbs and spices.

Technique: sous vide – food placed in plastic sachets, sealed under vacuum, cooked slowly then chilled.

Advantages: improves flavour of food, less shrinkage, suitable for most foods.

Technique: Chorleywood bread process
Addition of improvers and a few minutes of intense mixing thereby reducing time.

Technique: Modified Atmosphere Packaging (MAP)
Prolongs shelf life of fresh food; food can be kept up to 10 days instead of 2–3; meets demand for fresher products.

3. Wholemeal bread.

4. **(a)** **(i)** The percentage of the whole grain used in flour.
(ii) Flour improvers to improve texture; nutrients (iron, calcium, B-vitamin complex); bleaching agents (chlorine dioxide, soya flour); raising agents (self-raising flour).
(iii) Bran and wheatgerm contained in wholemeal flour weakens the gluten so a combination of white and wholemeal improves the texture. It also improves gluten content and can improve flavour.

(b) **(i)** A protein found in flour; made up of glutenin and gliadin.
(ii) When mixed with liquid becomes elastic and can be stretched; ability to stretch and hold pockets of air, gas produced by yeast; addition of salt and extra kneading, develops gluten and increases elasticity; when in oven it stretches as bubbles of carbon dioxide expand; in oven it coagulates to form a framework/structure for the baked item; allows dough to rise to give extra light texture.

(c) Use small equal-sized portions eg 50 g of flours of differing gluten contents:
(A) strong plain flour
(B) ordinary plain/soft/weak flour
(C) self-raising flour
Water to form a dough.
Place each dough in muslin or similar cloth.
Rinse out starch under gentle stream of water, squeezing bag to help.
Continue until no starch remains and water runs clear.
A small ball of gluten is left.
Weigh samples.
Work out percentage of gluten of different flours.
Conclusion: strong flour contains more gluten.
Gluten balls may be baked after rinsing.

5. INPUT
Information, materials, ingredients or energy that is put into the system (examples: eg yeast)

PROCESS

What happens to the materials as they go through the system; the changes that take place during making (examples: eg kneading, rising, proving, baking)

OUTPUT

What the system produces; the changes the systems bring about in the finished product (examples: eg finished bread product)

6. (a) Ensure accurate weighing of ingredients; use of scales/electronic equipment; equal division of dough into roll.

(b) In oven too long; oven temperature too high at first to kill yeast; then reduced to ensure that inside of dough is properly cooked; not rotated in oven.

(c) Insufficient fermentation/proving time; insufficient liquid resulting in a dough which is too stiff to allow expansion; inactive yeast which has not produced enough carbon dioxide.

(d) Sprinkling of poppy seeds/sesame seeds; glazing with beaten egg, etc.

7. (a)

Method of preservation	Suitable food product	Reason for choice
Freezing	Vegetables	Can be bought in bulk when cheap. Freeze well and cook like fresh. Nutritional value maintained. Bacteria are inactive at –18 °C.
Bottling	Marmalade/jam Fruits (peaches)	Air vacuum prevents recontamination. High sugar content prevents bacterial growth.
Canning	Tomatoes Beans/pasta Soups	Flavours maintained. Enzymes destroyed. Colour maintained. Air vacuum prevents recontamination.
Drying	Potatoes Milk Coffee Dried fruits	Reduced storage space. Convenient to have in cupboard. Lightweight. Lack of air and moisture prevents bacterial action.
AFD Accelerated freeze drying	Coffee Baby foods Herbs	Open texture when dehydrated quickly. When reconstituted flavour acceptable.
Pickling	Vegetables, eg onions Chutney	Acid kills micro-organism. Interesting flavours.

(b) Freezing occurs at approximately –10 °C therefore the water in the product forms into ice crystals in the individual cells. As tomatoes are very high in water content the cell structure is affected when thawing takes place. The loss of juices (or dripping) results in the tomato losing its structure and is therefore unsuitable to be served raw.

(c) (i) Milk is warmed then homogenized by forcing through a fine aperture to break up fat globules. Fat globules are then evenly suspended throughout the milk. Milk is heated to at least 132 °C. Milk held at this temperature for 1 second. All micro-organisms are destroyed. Cool rapidly and pack in foil-lined containers/sterilized containers.

(ii) Heating to 132 °C kills all micro-organisms. Homogenization ensures fat globules are evenly distributed throughout the milk.

(iii) Up to 10% loss of thiamine; 25% loss of vitamin C.

8. (a) (i) Re-heating/cooking facilities (with emphasis on safe reheating/cooking)
Portion control
Nutritional content (eg carbohydrate, low fat, protein, energy density)
Needs of sportsmen/women
Vegetarian ingredients
Cost
Dietary needs/healthy eating.

(ii) (Reasons must relate to point stated in **9 (a) (i)**
Re-heating facilities – microwave oven; suitable dish (disposal); speed; meals for one; safety probe/timing facilities
Portion size – suitable quantity; customer will be hungry; however not too large
Carbohydrate – provides energy; provides starch and sugars; breakdown of those nutrients
Low fat – reduce fat intake therefore reduce CHD therefore helps to keep weight down
Protein – provides some energy; growth and repair
Vegetarian – provides choice; should provide a protein alternative; provides vitamins, minerals, NSP
Cost – suitable price; competitive with other products
Healthy eating – recommended daily intake of food groups; helps reinforce government guideline.

(b) (i) Suitable choice must include a staple/carbohydrate food.

(ii) Provides energy through monosaccharides for all body activities; polysaccharides for long term energy; polysaccharides are cereals, pasta, rice; monosaccharides are glucose and fructose; dish will be filling because of staple food which provides bulk; relevant comments about protein foods (eg high biological value protein/amino acids) and iron for red cell production.

(iii) Product heated throughout; must stand after microwaving; no reheating of unsold products; underheating can cause food poisoning as foods must be re-heated to 70 °C or above for 2 minutes; use of probe to judge temperature accurately.

9. Both areas must be covered fully to gain full marks:

(a) Human expertise – selection of vegetables of high quality and good value for money; vegetables for uniformity/cleanliness/damage; possible during preparation of vegetable (depending on type) to avoid damage; checking; control over cooking of product; testing to ensure certain criteria are met, eg taste; costing of product.

(b) Robotics – possibly to check uniformity of vegetable size; weighing/measuring of ingredients; possibly preparation; cooking (temperature/time control); packaging.

10. (a) 2.25/2.2 g

(b) 12.5%

(c) Two ways to include:
increase vegetables; include fruit; use of wholemeal/granary four (explain fully).

(d) A clear understanding of the following effect is needed:
Fat in cheese melts and spreads, especially types such as Mozzarella; protein is denatured and can become tough and stringy due to overcoagulation of the protein and its separation from the fat and water; change in colour (golden brown) but depends on variety of cheese as some carbonize more quickly than others; fat content can become visible in some varieties giving an oily deposit on the surface; flavour and smell can become more pronounced; no significant effect on nutritional value although overhardening of protein can make it less digestible.

11. Answer must be related to fast food outlets:
Statutory requirement for all food handlers to be supervized and instructed and/or trained in food hygiene matters to a level appropriate to their job; greater financial penalties for breach of regulations; premises must comply with new regulations; equipment to comply with new regulations; recorded temperature control checks of refrigerator/frozen food storage areas; risk assessment to be carried out regularly; better working conditions for staff due to the design of premises (changing facilities, washing facilities etc.); staff to feel a part of the process in maintaining the regulations due to training and supervision throughout work; checking that the raw ingredients/materials meet the specification set; advertizing must be accurate and changes, without notification to customer, cannot be made; retail food prices have risen to accommodate the costs of implementing the Food Safety Act; public awareness of the responsibility the fast food outlet have to maintain the approved hygienic standards in the preparation and retailing of food products.

Sample coursework outlines

NEAB

Full course

Investigate how the consistency of starch-based sauces needs to be adjusted for products which are to be frozen. Using your findings, design and make a product which can be reheated after freezing.

Some people require special diets. Design and make products which could be sold by a major food chain in a special diet food section of its stores. Include appropriate labelling for the products.

The head chef of a catering company, which specializes in providing meals and buffets for special occasions, wishes to include menus for religious festivals. Design and make a range of food products which could be included in this new venture.

Short course

Investigate how different temperatures used during processing affect the appearance, flavour, palatability and nutritional value of vegetables. Use your findings to design and make an attractive vegetarian product.

Design and make a cake/biscuit which meets the consumer preference for a sweet product but which is actually low in sugar.

EDEXCEL LONDON

Full/short courses

Opportunities for design briefs			
Product development presentation	*Sports, pastimes and entertainment*	*Travel and environment*	*Diet and health*
New product development and specifications; Promotion and advertising of a product; Packaging; Corporate image	Food requirements for athletes; Party foods; Refreshments: packed lunches; Outdoor eating: picnics, barbecues	Cultural food; Availability of foods; Regional foods; Customs and traditions; Effects on environment, eg waste/pesticides; Environmental issues; Packaging	Individual nutritional needs; Effects of diet on health; Healthy eating/dietary goals; Nutritional content of prepared foods

OCR/MEG

Teenagers are amongst a growing group of people who eat a vegetarian diet. A manufacturer is to produce a food product for this group of people that represents a quick, low waste, low cost solution. You are asked to develop the product.

The market for between-meal snacks grows continually. Develop a novel savoury product that would be less damaging to health than the majority of choices on sale today that a real food company might produce.

The range of food offered at most buffet lunches is uninspiring, generally unappetizing and full of fat. Develop a new product to enable a catering company to change this image.

AQA/SEG

Sample coursework tasks are not provided by this board as they wish to keep the option of tasks as wide as possible. Students devise their own tasks but teachers can check their suitability with the board if necessary.

Index